*What we know about our star
and how we learned it*

Our Sun

Tony Broxton

*A journey exploring what we do know about the
Sun and what we have yet to learn, but also how
we discovered its secrets*

iUniverse, Inc.
Bloomington

Our Sun

iUniverse books may be ordered through booksellers or by contacting:

iUniverse
1663 Liberty Drive
Bloomington, IN 47403
www.iuniverse.com
1-800-Authors (1-800-288-4677)

ISBN: 978-1-4620-5570-8 (sc)

Printed in the United States of America

iUniverse rev. date: 10/11/2011

Disclaimer

The author has taken all reasonable care to ensure that all data contained within this publication is accurate on the stated date of publication or last modification. The author has taken all reasonable care to comply with the copyright, intellectual property rights and licence agreements of sources from which material has been used in the preparation of this book.

It is in the nature of websites, many of which are subject to change, that information published may be inaccurate, out of date, for test purposes only or the personal opinion of the author. Readers should verify the information gained from the Web with the appropriate authorities before relying on it.

Author's Personal Profile

I am a British professional chartered electronics engineer and physicist, and started my career as a lecturer in physics, mathematics and electronics. I then became a designer of aircraft avionics, specialising in aircraft electronic landing systems, later becoming a member of the Tornado aircraft design management team.

I joined British Aerospace where I worked for over 22 years on various aircraft and missile systems. My work involved tours of duty at the Lockheed Missile and Space Center, (LMSC), Sunnyvale, California and at the Atomic Weapons Establishment, Aldermaston, England. During my career with British Aerospace, I was the head of a weapons forensic laboratory for 11 years.

I rose to the position of British Aerospace Consultant Engineer, their highest engineering grade, as a specialist in the field of semiconductor technology. As a Consultant, I undertook three years research at the University of Maryland, in consortium partnership with the UK Ministry of Defence, the US Department of Defense, and the NASA Mars Pathfinder project team, amongst others. I also came into professional contact with the NASA Galileo team, involved in the Jupiter missions.

Aged 53, I retired from British Aerospace in 1998. Working with NASA scientists and engineers rekindled a boyhood interest in the subject and so I retrained to become an astronomer. I was invited to join a local amateur astronomical research team where I started undertaking solar observational work, which I submitted to the British Astronomical Association, (BAA). The BAA subsequently offered me the post of Assistant Director (Projects), of the Solar Section, a position held for three years. In 2006, I was elected a Fellow of the Royal Astronomical Society.

I live in Cornwall, England, with my wife Barbara where I still undertake daily solar observational work (weather permitting) as part of a 70 strong BAA international team, and am a member of the international Association of Lunar and Planetary Observers (ALPO). I report to these organisations monthly. I submit my work daily to the Solar Influences Data Analysis Center (SIDC), in Brussels. The SIDC is the world centre of the Regional Warning Center (operating through an NOAA hub in Boulder, Colorado) which issues solar storm warnings to airline companies, satellite operators and other interested organisations.

Other books: I am also the author of the *"Solar Observer's Handbook"*, ISBN 978-1-4389-1140-3(sc), written both for the novice and more advanced astronomer, as well as amateur astronomy groups. The book covers the basics of the science, taking extra steps to follow through to a level that will enable the reader, or an astronomy group, to submit their own work at a national level, should they choose to, thereby enhancing their enjoyment of the subject, and by making a genuine contribution for the benefit of all.

Note.

Whenever a term appears for the first time in his book, it is printed in italics, accompanied with an explanation. Some terms have been embellished with additional information in the *"Glossary of Terms"*.

------------------*Contents*------------------

Introduction

Open any book on the Sun and a bewildering array of facts are presented about a place that no one has ever visited, or is ever likely to. As no probes or satellites have gone there, how did astronomers get all their information? To make matters worse, one book may well give a value for some parameter or other that differs from that in another book. The reason for this is that over recent years there has been an avalanche of discoveries, with technological advances in measuring instruments and methods that have enabled us to refine our knowledge. Today there is a veritable armada of satellites orbiting our planet with their eyes on our Sun monitoring its behaviour and unravelling its mysteries. This book will not simply provide the facts, as are known at the time of writing, but will show just how these facts were obtained, or arrived at, and how the measurements were conducted on this object, which is far beyond our reach.

One may ask, *"how do we know what the internal construction of the Sun is?"* On the other hand, *"how do we know how the Sun actually works?"* The simple answer is *"we don't"*. We do know how materials, chemical elements, matter and the laws of physics behave on Earth and scientists extrapolate this data and perform calculations to try to account for the Sun's characteristics from those facts that we can determine from observation and experiment. It is by these methods that we claim to be able to explain many of the Sun's mysteries. Unless someone, or something, actually goes there and gets first-hand evidence, we cannot be completely sure, and that quite simply is just never going to happen. Therefore, in nearly all cases, when some fact or figure is given about some solar attribute, one must always leave the door open to the possibility of future change. Although we believe we know quite a lot about the Sun, there is still a great deal it has yet to teach us.

For example, when the first attempt to find out what the Sun was made of, back in 1814 by the German scientist Joseph Von Fraunhofer, he managed to identify nine chemical elements. The modern number given by researchers in 2009 stood at 69 elements and as instruments become ever more sophisticated, capable of detecting even smaller trace quantities, this number may well increase.

We know, for example, that it is millions of miles away, and hot enough to melt anything that approaches it, and that heat and light from it takes a little over eight minutes to reach us. It is also our nearest star, our next nearest neighbour being Proxima Centauri, and light from there takes about 4 years and 3 months to reach us.

It has taken us just a few generations to acquire this knowledge, but since the dawn of time, ever since Man relinquished his nomadic hunter lifestyle, settling down in one place and starting to farm crops, he took a greater interest in the Sun. He learned to understand the Seasons and to predict the best times to plant seeds and harvest his crops. He already knew that the summer brought warmth and growth and the Sun rose higher in the sky and the days were longer. The winter was cold, the trees bare, and the Sun hung low in the sky. Moreover, throughout the year the pattern of stars moved across the heavens to return once more at the same time the following year. By observing and noting what star patterns, which we now call constellations, appeared and when, coupled with the development of mathematics, an annual calendar could be devised from which he could organise his existence to his best advantage. These things passed into general knowledge through simple observation, as evidenced by the presence of a number of Neolithic stone circles constructed some 5000 years ago as primitive observatories in Europe and the British Isles. These stone circles predate the Pyramids, themselves believed to be linked to astronomical features such as the three stars that form Orion's Belt. Nearly 4000 years ago, by noting this "cycle of life", the ancient Babylonians believed a year was 360 days long and from this, they devised the sexagesimal or base-60 system, from which originated 360 degrees in a circle, 60 minutes in an hour, and so on. The system formed the basis for our present calendar, which would have consisted of 12 months of 30 days each. Since 60 is divisible by a significant number of integers, namely 2, 3, 4, 5, 6, 10, 12, 15, 20 and 30, it was considered special. This, then, was the birth of both mathematics and astronomy.

The economy and prosperity of ancient Egypt was umbilically linked to the annual flooding of the Nile. Predicting the start of the season was crucially important to the region's agriculture. Through millennia of astronomical observations, the Pharaonic Egyptians devised a calendar numbering 365 days. It was comprised of twelve months, with each month consisting of three ten day weeks, called *decans*. Five additional holy days were included, festivals dedicated to the worship of the gods. The New Year started on July 19 or 20, and every 1460 days, (or four years), an extra day would be added in the same manner as we now include an extra day for Leap Years.

But how do we know all this? The answer is the *Dendera Zodiac*. About 70 kilometres, or 45 miles, North of Luxor, known in antiquity as *Thebes*, is a township on the west bank of the Nile named Dendera. The ruins of a Greco-Roman temple complex are to be found there, dedicated to the Egyptian goddess of fertility, *Hathor*, which dates to about 50BC. Inside is a chapel dedicated to Osiris and in the ceiling was a carved sandstone zodiac, shown in the engraving below.

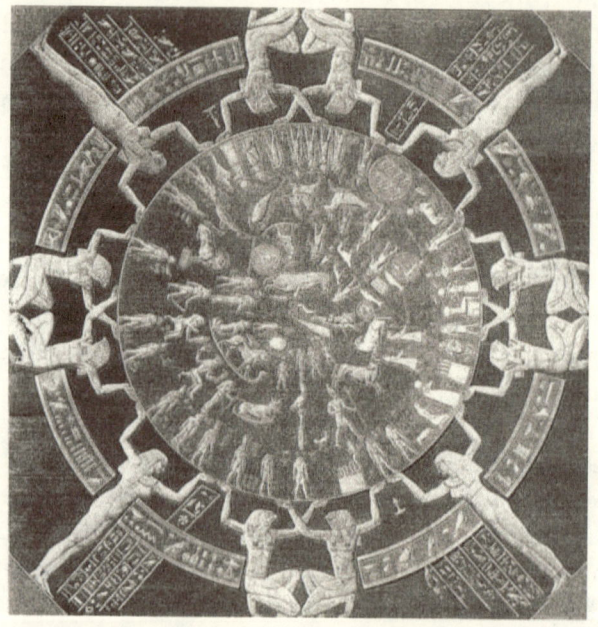

Figure 1   THE DENDERA ZODIAC

This zodiac, which is now in the Louvre Museum in Paris, provides all the details of the ancient Egyptian calendar. In addition, it shows the *heliacal rising* of the star *Sirius*. Its name in Egyptian was *Sopdet*, and *Sothis* in Greek. Sirius is the brightest star in the heavens and is aligned to the three stars forming the belt of Orion, which themselves held some religious significance for the ancients. Heliacal rising is a state where a star becomes visible above the Eastern horizon just prior to sunrise, following a period when it previously had not been visible. This occurs at the same time every year, around June 10 in antiquity heralding the start of the Nile floods.

Man has always been curious about nature and the world in which he lives, and the strange mysteries of the heavens above him. He would naturally be prompted to muse over such things as how far away the Sun is, and how big it is. Since it provides heat and light, what temperature is it? It was well known that it gets hot if you sit too close to a campfire and as you move away from it the air around cools down, which suggests that if the Sun is a long way away, it must be very hot indeed.

From simple observation, prehistoric man may have harboured the thoughts that the Earth was flat. Even today, there are those who believe the World is flat, supported by a giant turtle, even in the face of crushing scientific evidence. Popular belief that in medieval times the World was generally considered to be flat, is due in no small part to a work of fiction by Washington Irving entitled *"The Life and Voyages of Christopher Columbus"*, first published in 1828. Some, at the time, believed it to be a factual work, and so the myth was promulgated. More recently, Samuel Rowbotham (1816 to 1884) proposed that the Earth is indeed flat based on his interpretations of certain biblical passages. This led, in part, to the inception of *"The Flat Earth Society"* by Samuel Shenton in 1956. Certainly, the ancient Greeks were not convinced of this. They well knew that one could see the sail of ship coming over the horizon, long before the hull became visible. That had to mean that in the very least, the Earth was curved. Aristarchus from the Greek island Samos wrote in the 3rd century BC *"that the Earth is a ball and it spins."* Eratosthenes (276 to 195 BC) certainly believed the Earth to be round since he calculated its circumference to be 25000 miles. Around the equator, it is actually 24902 miles and 24860 miles around the poles, so his results are remarkably good for those times. This he calculated from noticing that in Egypt at the *Summer Solstice*, (when the Sun is at its annual highest point in the sky), the Sun shone directly down a well at Aswan. On the same day of the year, a stick placed in the ground at Alexandria cast a shadow equivalent to $1/50^{th}$ of a circle, or at an angle of 7° 12′ in modern values. Knowing the distance between the two sites, one can calculate the answer.

Then there are the *auroras*.

Charged particles from the Sun ionise the nitrogen and oxygen gases high up in our polar atmosphere. These were not immediately attributed to be an influence originating from our Sun, but were considered by some to be a manifestation of the gods and by some Indian tribes to be the spirits of the departed. Auroras have also been seen to exist on other planets in our Solar System, namely Jupiter and Saturn.

Their natural beauty hides a darker side. We now know that these aerial displays also bring radiation, which can damage the health of airline crews making repeated flights over the poles. Airlines have to make flight adjustments two or three times a month on average, and fly at lower latitudes, adding millions of dollars to operating costs. One flight change made by United Airlines to avoid a particular solar storm affecting our North Pole required an additional 3,000 gallons of fuel and the change cost the airline $10,000, more than the initial cost to operate the flight in the first place. We have also experienced this *solar wind* as it is called; induce massive electric currents in long metallic structures such as pipelines, railway lines, telephone and power lines, and even in the girders of high-rise buildings and oil rigs. Such events have lead to power black outs. One notable geomagnetic storm took place in March 1989, knocking out the Quebec region of the Canadian national power grid, and other power stations in the north-eastern United States. In the recent past, it has destroyed satellites. It has been estimated that damage to satellites due to solar storms have cost some $4 billion US dollars. Modern satellites such as those used for global positioning and mobile telephones employ design features to protect them from this damage. Yet even now, it still has the capacity to push them out of their designated orbit, necessitating periodic repositioning. It is a constant threat to astronauts.

Solar storms, *flares* and other eruptions on the Sun's surface cause surges in this solar wind that can blackout long range radio communications and radar.

Therefore, the Sun affects our daily lives, our way of life and our very health. Indeed, the changing solar cycle alternating between periods of both high and low activity has been linked to the changing price of wheat through its influence on our weather. It is important, therefore that we make the effort to learn as much about it as we can, and its behaviour, so as to predict its moods. We cannot control it and we never will. By understanding it, we can capitalise on its beneficence and protect ourselves from its malevolence. Starting with simple observations and learning the basic facts, we can now explore how the sciences went about extending our knowledge and unravelled the mysteries of our very own star. During this quest, we have also learnt that the Sun itself has a limited life and one day, in its death throes, it will finally destroy our own home planet, Earth!

Warning

Most books on the subject will carry a warning aimed at those who wish to make observations of the Sun, and this book is no different. Readers may well be tempted, indeed encouraged, to embark on the hobby of solar astronomy themselves, in which case they will need to exercise extreme caution.

It is very dangerous to look at the Sun directly, even with the naked eye, as it could seriously damage your eyesight. Sunglasses often provide no protection whatsoever.

Under no circumstances should anyone ever view the Sun through binoculars, a magnifying glass, telescope, or any optical instrument, as this could cause immediate blindness!

The Sun's rays focussed through a magnifying glass can be used to ignite a piece of paper or even wood. If the Sun is viewed through any optical instrument, the Sun's rays can burn out the optic nerve causing immediate and incurable blindness.

Astronomers always use telescopes specifically designed for the job, fitted with special solar filters that remove the harmful ultraviolet and infrared light that would otherwise blind them.

The Sun's rays can be strong enough to cause damage to one's eyesight even if viewed directly with the naked eye,

Never look directly at the Sun!

The Life Cycle of Stars

To better understand the following chapters, and in particular *"How Old is the Sun?"*, and *"What is the Sun Made Of?"* it is worth examining the manner in which stars are born, evolve, journey through their lives and eventually die. New discoveries are continually being made which challenge these theories, but this chapter outlines the process as is currently believed to be the case.

The story of the life cycle of stars has been deduced from countless observations of the Heavens coupled with some laboratory experiments performed here on Earth employing the known laws of Nature. However, it is still a subject of much study and research. It starts in the vast expanse of space itself, a void that is not a perfect vacuum, but actually contains occasional atoms of hydrogen gas, and sub atomic charged particles, namely *protons* and *electrons*. An electron is a negatively charged particle and a proton is a positively charged particle. The two electrostatic charges are of equal magnitude, but these particles have vastly different masses. A proton has 1836 times the mass of an electron. Hydrogen is the simplest atom, comprised of just one proton and one electron.

Intergalactic space is the void between the galaxies. Interstellar space is that between the stars within galaxies including our own galaxy, the Milky Way. Here there will be a slightly higher occurrence of hydrogen atoms than in the far greater expanses of intergalactic space. Measurements made by deep space probes, albeit within our own Solar System, indicate that there are generally only a few atoms of hydrogen gas per cubic metre, occasionally rising to as many as 40 atoms per cubic metre. By comparison, the air we breathe on Earth contains around 10^{25} atoms, or 10 trillion trillion atoms in every cubic metre. Interstellar gas clouds, called *nebulae*, can occupy vast regions of space, be light years across, and although they appear to be quite substantial, in reality they generally have an extremely low gas density, often lower than the hardest vacuums we can produce on Earth.

In 1687, Sir Isaac Newton published his *"Principia Mathematica"* and in it he described his famous Laws of Gravitation. He expounded that all bodies in the Universe, from whole galaxies right down to the humble atom, were gravitationally attracted to each other in a manner dependent on their individual mass and the distance separating them. This is dealt with in further detail in the chapter *"How Big is the Sun?"*

Now in the vast reaches of space and over millions of years, individual atoms of hydrogen, and any other matter, would become gravitationally attracted to each other and eventually join company. This process will continue and as more atoms join the cluster, their combined mass increases pulling in further unattached atoms, charged particles and any space debris (dust) from the surrounding neighbourhood towards itself in a seemingly "runaway" affair. Gas clouds eventually form to become nebulae in this basic way. There is a second way in which a nebula can form, discussed later in this book. The gas pressure and density distribution in a nebula will not be regular, and there will be regions where gas densities will be slightly higher than the norm. Material in these locations within the nebula will coalesce into even higher density regions in a random manner, gathering material, atoms and charged particles, from the surrounding regions. As these centres develop, and their gas pressures and densities rise above that of their immediate environs, due to their higher gravitational pull, they will rob the neighbouring areas of gas and any other material that might be present. First proposed by Sir James Jeans towards the middle of the 20th century and known as Jeans' Instability Criteria or Jeans' Criteria for the Collapse of a Nebula, by employing the known laws of physics, this process can be mathematically modelled. There are computer simulations embodying Jeans' formulae, which manipulate random computer generated nebula graphics that can dynamically demonstrate the formation and evolution of stars and planetary systems.

The process now takes on another twist. As the denser regions continue growing, they become denser still, and start collapsing in on themselves under the influence of their own gravity. If the process is allowed to continue, if it has sufficient original mass and material, it will start to heat up, as the atoms comprising it collide with each other with ever-increasing frequency and velocity. When the temperature reaches about 10 million degrees this ball of gas will start emitting light, and a star is born! At this stage, it is called a *protostar*. Contracting masses of hydrogen gas as low as 8.5% that of the mass of our own Sun are sufficient to generate temperatures of this magnitude. Observations have shown that within a nebula, there may be many stars forming, and nebulae are often called the birthplace of stars, or stellar nurseries. This is one reason why some stars tend to form in clusters.

The process continues. Gravity draws the hydrogen gas in on itself and the temperatures rise even more. The process reaches a balance point where the forces of gravity match the force of expansion from the hot gases. The exact point at which this occurs depends on just how great a mass of hydrogen gas, and any other material, is involved. Since the forces of gravity concentrate in the centre of mass, so does the temperature, and there will be a temperature gradient with the inner regions hotter than the surface. The colour of the emitted light will also change. Some stars appear to emit white light, others bluish, yellow, orange, or some an apparent reddish hue. There is a direct relationship between these colours and their temperatures, and this is dealt with in detail in the chapter on *"How Hot is the Sun?"* The surface temperatures of the white or bluish stars are reckoned to be as high as 80,000 degrees, decreasing to around 2,500 degrees in the case of the reddish stars. The inner regions, or the cores of these stars, become so hot, reaching many millions of degrees, that they become nuclear furnaces. Called *stellar nucleosynthesis*, chemical elements are created by nuclear fusion, a process in which sub atomic particles combine and release energy. This assumption is based, once again, on the known laws of nuclear physics and chemistry, and by calculation, as well as spectroscopic analysis of the light received from the stars. Details of this spectroscopic analysis are covered later in this book.

Various processes are believed to occur. The first of these is called the *proton-proton chain* or sometimes simply the "p-p chain" for short. Here helium is created, the second commonest element in the Universe. It works like this.

Two protons collide and bond to create a *deuteron*, which is a proton and *neutron* pair. The neutron has nearly the same mass as a proton but has no charge. An electron antiparticle, called a *positron* is emitted, carrying off the positive electrostatic charge previously carried by one of the protons. Another subatomic particle, a *neutrino*, is also emitted.

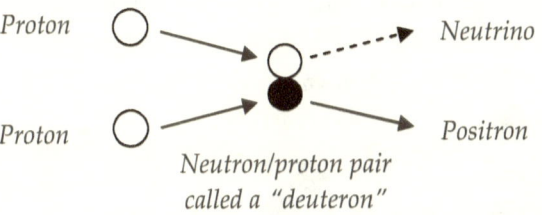

Figure 2   INITIAL PHASE OF THE PROTON-PROTON CHAIN

Here the deuteron is the *nucleus* of *deuterium*, or heavy hydrogen. Since it is in a transient phase in the process, it is not to be confused with a *nucleon*, the name given to a bound neutron/proton pair that resides in the nucleii of atoms. All other atoms contain neutrons as part of their structure with the sole exception of hydrogen. The second part of the process involves the bonding of another (third) free proton with the deuteron to form an isotope of helium called *helium-3*, and results in the emission of electromagnetic energy in the form of a *gamma ray*.

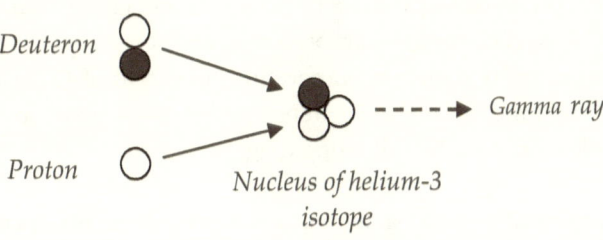

Figure 3   SECOND PHASE OF THE PROTON-PROTON CHAIN

The third part of this story involves the fusion of two of these helium-3 isotope nuclei to form the nucleus of a helium atom, with the emission of two protons, which can go on to either start the whole process all over again, or take a role in further reactions. The helium nucleus is also known as an *alpha particle*.

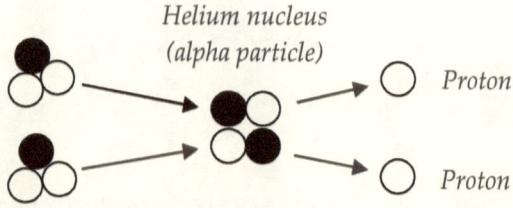

Helium nucleus (alpha particle)

Proton

Proton

Figure 4   THIRD PHASE OF THE PROTON-PROTON CHAIN

In the fourth and final phase, the helium nucleus combines with two free electrons to become a helium atom.

However, there is more to the story. The positron emitted in the first part of the process may encounter an electron, which has an equal and opposite electrostatic charge, whereupon they will annihilate each other producing more gamma ray energy.

A second process can produce both helium and energy, known as the *C-N-O cycle,* or the Carbon-Nitrogen-Oxygen cycle. This process generally requires temperatures in excess of 15 million degrees to function. The sequence of events is given in the table below, and in the following diagram.

1) Carbon 12 combines with a proton to form the isotope nitrogen 13, releasing gamma ray energy.

2) Nitrogen 13 decays into carbon 13, releasing an electron and a neutrino.

3) Carbon 13 isotope combines with a proton forming the isotope nitrogen 14, also releasing gamma ray radiation.

4) Nitrogen 14 combines with a proton to form the oxygen 15 isotope, again releasing gamma ray radiation.

5) Oxygen 15 decays into the isotope nitrogen 15, releasing an electron and a neutrino.

6) Finally, nitrogen 15 combines with a proton to form the original carbon 12 once again and releases a helium atom nucleus, or alpha particle.

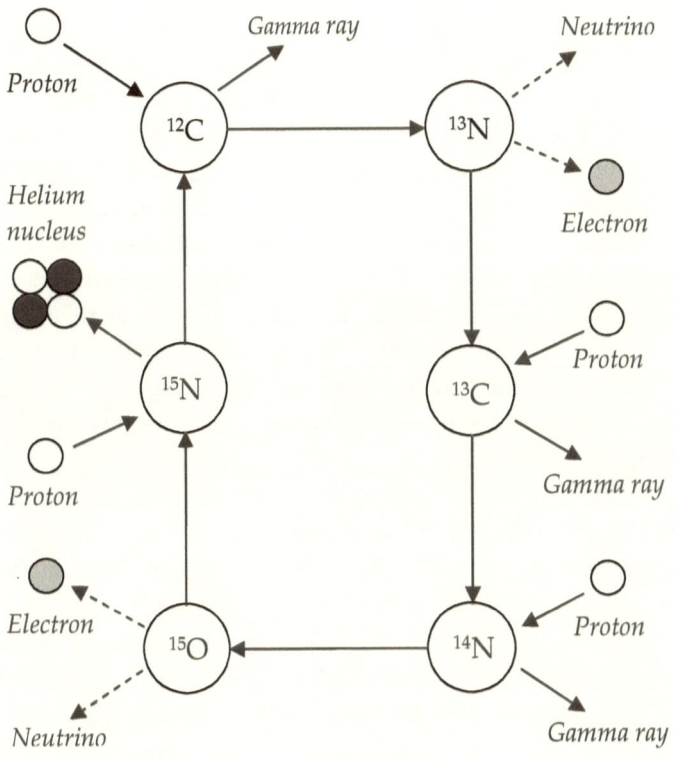

Figure 5   THE C-N-O CYCLE

These processes are known to occur from spectroscopic analysis of the light received from the Sun. All of these isotopes have been detected and so are known to exist. The *atomic mass* of an isotope, as shown in the previous diagram, defines the number of neutrons and protons present in the nucleus. The *atomic number* of a chemical element simply defines the number of protons in its nucleus. For example, oxygen has an atomic number 8, its nucleus containing 8 protons. The basic atom will also have 8 neutrons in its nucleus and be designated with the atomic mass Oxygen 16. The isotope Oxygen 15 still has 8 protons in its nucleus, but now has only 7 neutrons.

If a star's core temperature is high enough this process of nuclear fusion may continue, next creating the elements beryllium and lithium via a similar process, and thereafter further beryllium plus boron, although these more advanced processes produce less energy. Chemical elements up to atomic number 26, iron, can be produced by nuclear fusion, a process that releases energy. Thereafter, heavier elements will need to be created through a process called nuclear fission, a process requiring the consumption of energy to function. Therefore, within the nuclear furnaces of stars like our Sun, chemical elements can be created up to iron. It is possible that transient hot spots may occur within the core furnace allowing the creation of higher order elements such as cobalt and nickel, atomic numbers 27 and 28, but this is conjecture. Higher order elements, if they are generated, though no one knows for certain, risk being broken down into simpler elements. However, much heavier elements do exist in the Sun and stars, and how these are formed will be detailed later.

Now while the gas cloud that will eventually form our Sun is contracting, it may not have been perfectly spherical and its centre of gravity may not have been at its epicentre. There may well have been within it, regions of relatively higher density material compared to its immediate surroundings. Two or more such regions interacting gravitationally with each other will cause the system to start to rotate, to spin about its true centre of gravity, although there need not necessarily be any particular structure present at its apparent centre of gravity. Concentrations of high-density gas on the periphery may drift off under the influence of the centrifugal forces present. They themselves will spin, generally in the same direction and plane, about the central body. The central body will eventually go on to become a star, if it has enough material, and the outer satellites, the smaller spheres of gas, may either go on to cool and eventually become the planets, or if they have sufficient mass, go on to become stars in their own right. This is how binary star systems form. Thus, the planets in our Solar System orbit the Sun more or less in the same plane, which is called the *Plane of the Ecliptic.* The Sun and the planets will tend to rotate in the same direction also, unless they collide with another celestial body or suffer some other trauma. The Sun and the Earth both rotate in an anticlockwise direction when viewed from above, looking down on their respective North poles.

Having separated from the main ball of collapsing and heating gas, for planets to form around the star, their temperatures will start to fall due to them not having sufficient mass and gravity. Some solidify like our four inner planets Mercury, Venus, Earth and Mars whilst others like our four outer planets Jupiter, Saturn, Uranus and Neptune remain in a cold gaseous state. Though outwardly cold, they may still have a hot core, but not hot enough for nuclear processes to occur. In the same manner, whilst in their embryonic gaseous state, the forming planets might also throw out parcels of cooling gas from themselves to become moons. In our own system, some planets have captured asteroids in their gravitational fields in addition to giving natural birth to daughter moons.

The stars and planets, having formed, now set off on their journey through life. This journey will depend on their size and temperature. The process is depicted in a diagram known as the *Hertzprung-Russell Diagram*, and a simplified version is shown here. The Hertzprung – Russell diagram or HR diagram for short, cross relates the star's surface temperature, determined by its colour, against its luminosity or brightness *relative* to that of the Sun. The relationship between a star's surface temperature and its colour is explained in the chapter *"How Hot is the Sun?"* The *Spectral Types* shown below can be subdivided further into ten subgroups, and it can be seen in the diagram that our Sun is designated as a spectral type G2.

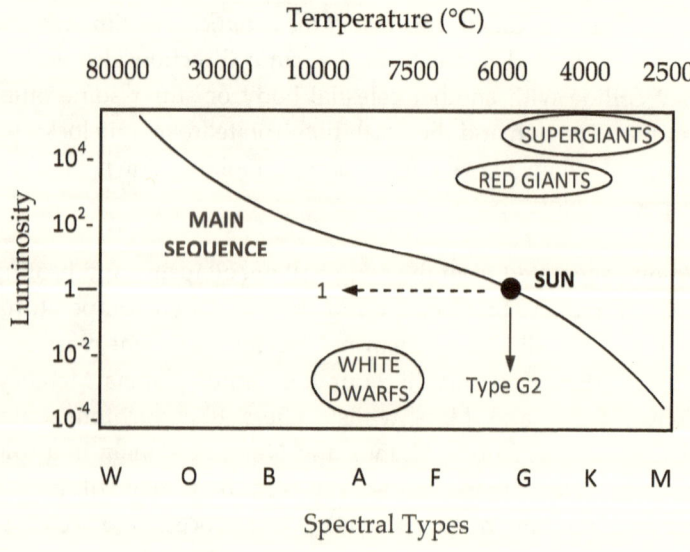

Figure 6   SIMPLIFIED HERTZPRUNG–RUSSELL DIAGRAM

The chart shows the domain of just a couple of star categories, red giants and white dwarfs, as it is not possible to show all of them here. The different colours/temperatures of stars are categorized as Spectral Types according to the following table. At one time there was a sequential sequence running from A to O. The classification of Spectral Types was reordered into the present listing once the relationship between colour and temperature was realised. Spectral type "W" was added to identify the brilliant white category.

To remember the sequence, some astronomers use the mnemonic *"Wow, Oh Be A Fine Girl/Guy, Kiss Me"*!

Spectral Type	Surface Temperature, (°C)	Colour
W	Up to 80,000	Brilliant white
O	50,000 to 30,000	Blue
B	30,000 to 11,000	Blue – white
A	11,000 to 7,500	White
F	7,500 to 6,000	Yellow – white
G	6,000 to 5,000	Yellow
K	5,000 to 3,500	Orange
M	3,500 to 2,500	Red

Stars with insufficient mass will simply burn themselves out in a relatively short period of (astronomical) time. Once a protostar has formed which is of sufficient mass to condense and become a fully fledged star, it will take its place on the main sequence curve, at a point dependent on its temperature. Here it will spend the best part of its life converting hydrogen into helium, and radiating energy, gases and charged particles from its surface. This process may last for billions of years. As time advances, the star shrinks and cools, and its position on the Main Sequence curve changes accordingly. Helium builds up in the core whilst the hydrogen is consumed until it runs out altogether. There will be a conflict between the forces of gravity, which attempt to pull the star in on itself, and the internal temperatures, which try to make the star expand. Initially gravity wins and the star contracts, but in so doing the temperature rises once again. Now the helium in the core begins to create heavier chemical elements, causing the inner regions of the star to contract further, whilst the outer layers expand. A point will eventually be reached when one becomes the victor. This is the point in time when the star starts to die, and its final swan song depends on the mass of the stellar corpse.

For ease, astronomers relate the mass of a star to that of the Sun, the Sun being taken as one *solar mass, M_0*. A star, such as our Sun with a solar mass M_0 of one, can expect to reside in the Main Sequence for a period of about 10 billion years. Larger stars with solar masses M_0 around ten, can expect to spend only a few million years in the Main Sequence. The time spent by a star in the Main Sequence is an estimate based on a calculation surrounding its temperature, mass, and hence volume and surface area, providing a figure for the rate at which it burns up its nuclear fuel. They will then follow either of two different paths when they exit from the Main Sequence, determined by their terminal masses, shown in the diagram on the following page.

Some 70% of stars in our galaxy are binary stars. These are two or more stars rotating about each other. By measuring their rotational parameters and using Newton's Laws of Gravity and Kepler's Laws, covered in the chapter *"How Big is the Sun?"* their stellar masses can be determined. Most stars fall within the range of one to a few tens M_0.

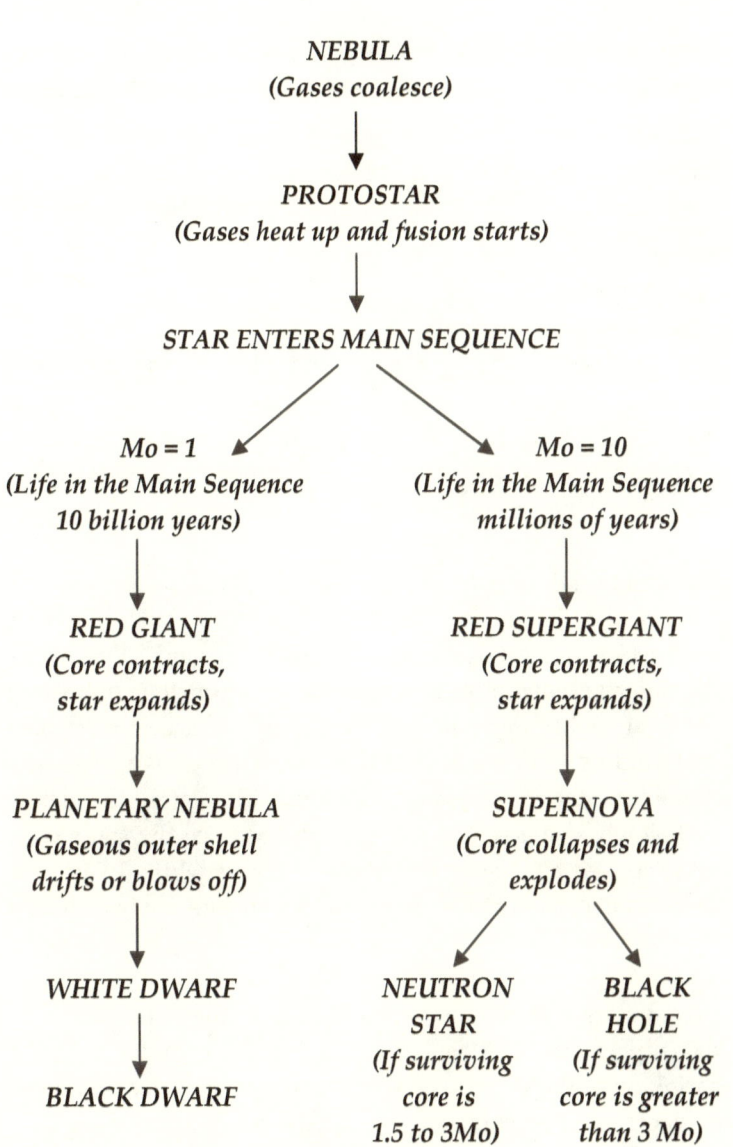

Figure 7   THE LIFE CYCLE OF STARS

Catalogued as R136a1, the largest star listed at the time of writing is a hypergiant, estimated to be about $320M_0$ at birth, but now around $265M_0$. It was discovered in the Tarantula Nebula, in the constellation Dorado, part of the Large Magellanic Cloud. It is thought that the minimum mass that a star can have lies somewhere between $0.04M_0$ to $0.1M_0$, though opinions differ.

For stars with solar masses around $M_0 = 1$, the process is believed to be as follows. The star cools and expands becoming a red giant. In the future, our Sun will expand, becoming a red giant and envelop the inner planets, possibly including the Earth.

Eventually all the available hydrogen gets converted into helium and the nuclear processes begins to stall. The core contracts and the outer atmosphere will expand. Eventually the outer gaseous atmospheres either drift off the contracting core or will explode off, creating a gaseous nebula surrounding the remnants of the core, which will appear as a tiny white star called a *planetary nebula*. Thereafter, the nebula will slowly disperse and leave what was once the star's core, now labelled a *white dwarf*. The fate of this white dwarf is to cool and eventually fade out altogether, eventually becoming a *black dwarf*.

Larger stars will cool and expand in the same way becoming red supergiants, but will take far less time to do so. The shrinking of the larger core will create temperatures of billions of degrees. The core will eventually become predominately iron, which will not react further and the nuclear processes cease. At this point the core implodes and immediately thereafter explodes in a cataclysmic fashion. This explosion is called a supernova. The energy in a supernova is the only known energy source in the Universe high enough to create all the heavier chemical elements, right through the transuranic range, by bombarding atomic nucleii with neutrons and protons through a process called *supernova nucleosynthesis*.

Much of the star's material is blown off into space where it can seed other nebulae with its chemical treasure trove. Where the mass is lower than $3M_0$, there remains an incredibly dense, hot and very small core composed mostly of neutrons, possibly with some electrons in the intervening spaces, created by the enforced combination of protons and electrons by the force of gravity. This is a *neutron star*, which may be only a few kilometres in diameter, but it will have a massive magnetic field. Some rotating neutron stars named *pulsars*, can spin very fast, often several hundred times a second, and have the property of emitting radio beams. Discovered by NASA in October 1999, at a distance of 39,000 light years in the constellation of Ophiuchus, is a neutron star labelled XTE J1739-285 that has been seen to rotate at 1,122 times a second.

Dead cores with masses greater than $3M_0$ contract under the forces of gravity in a runaway manner whereby, as they contract, gravity increases causing them to contract still further. Their density and internal gravitational forces rise to such unimaginable levels that even light itself cannot escape their clutches. These are called *black holes*.

The Structure of the Sun

We know the Sun is a spinning ball of hot gas radiating heat and light, and a solar wind comprising a *plasma* of ions, electrons and protons, together with electromagnetic radiation spanning the entire spectrum from radio waves to x-rays and gamma rays. Internally it is composed of various regions, and even has a rudimentary atmosphere, called the *chromosphere*, and an extensive very thin outer gaseous region, the *corona*.

In order to understand the following chapters better, it would help to have a basic understanding of how we believe the Sun is constructed. Below is a diagram showing the various internal regions of the Sun.

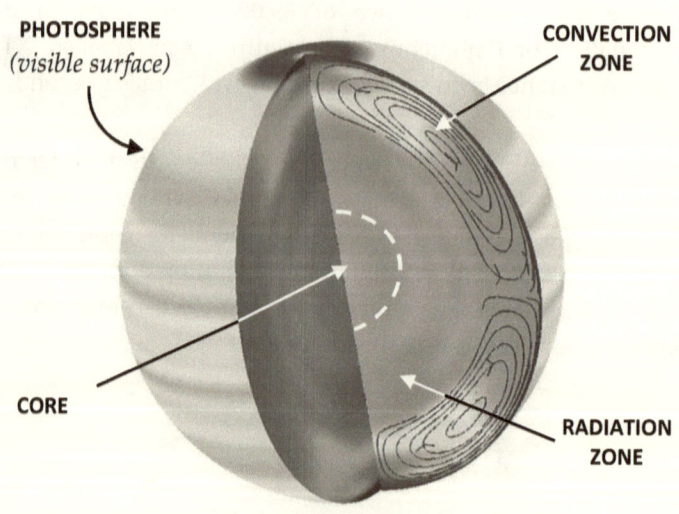

PHOTOSPHERE
(visible surface)

CONVECTION ZONE

CORE

RADIATION ZONE

Figure 8   THE CONSTRUCTION OF THE SUN

So, how did astronomers produce this model of the Sun's interior? The present model was developed using a branch of science called *helioseismology*.

The Earth science of seismology, or geoseismology, is used to map the rock strata beneath the Earth's surface, and to prospect for minerals and oil. In early days, the shock waves from earthquakes were monitored and analysed to try to determine the subterranean geology. In the first half of the 19th century, explosive charges were used to induce shock waves into the ground and echoes reflected off the various subterranean features were detected, recorded and analysed, enabling detailed maps to be drawn up.

Similar techniques are used to map the Sun's interior. Obviously, shock waves cannot be induced into the Sun by human intervention. However, it is not necessary. Observation revealed that the Sun wobbled and pulsated slightly like jelly, due to the creation of acoustic pressure waves generated from the turmoil within its interior.

These acoustic pressure waves originate from several sources, and are very similar to sound waves but have very long wavelengths outside the human hearing range. They range in periodicity between 1½ to 20 minutes. As the Sun pulsates these pressure waves cause slight changes in the wavelengths (Doppler shifts) of the absorption spectrum lines emanating from the Sun's surface. Satellites make these measurements.

The source of these pressure waves can be identified by analysing these Doppler shifts. By coupling this information with knowledge of the Sun's mass, overall density, luminosity, temperature, and chemical composition, it is possible for scientists to develop a model of how they think the interior of the Sun is configured. However, to complete this model, astronomers had to make assumptions about a region called the convection zone, which is covered later. Therefore, the accuracy of this model cannot be verified and probably never will be, but currently it is the best guess there is.

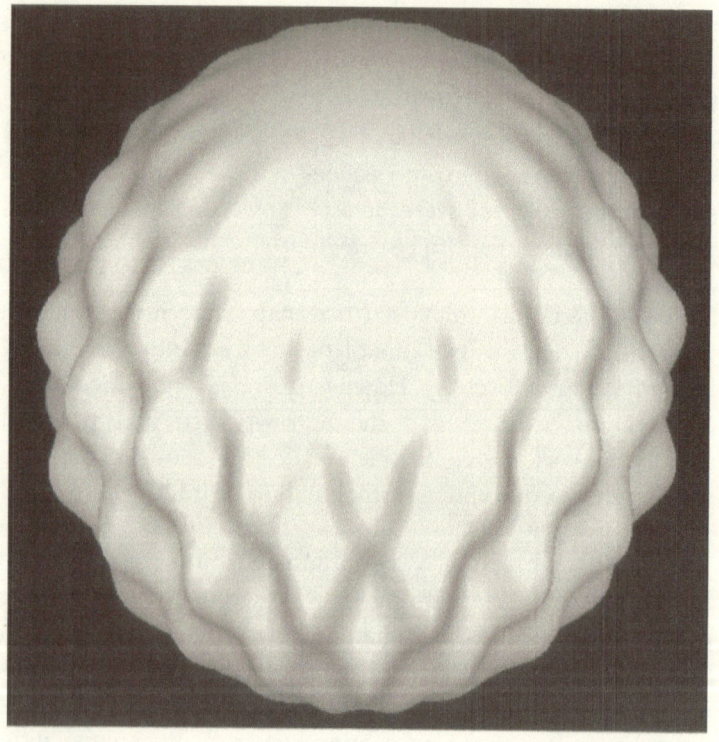

Figure 9   ARTIST'S IMPRESSION OF SOLAR ACOUSTO-SEISMIC OSCILLATIONS, HIGHLY EXAGGERATED

The image above, reproduced here with the kind permission of the NASA Marshall Space Flight Center, graphically illustrates this principle. The facts and figures on the structure of the Sun given hereafter originate from these helioseismic models.

The *photosphere* is the surface of the Sun that we can actually see. It is gaseous and is not a fixed solid surface in any sense of the word. The temperature of the photosphere is on average 5500°C or 5776K, (where K = degrees Kelvin), as explained in the chapter on *"How Hot is the Sun?"* It is a region somewhere between 200 kilometres and 500 kilometres thick.

Seen through a "solar" telescope it appears as an ever changing cellular surface, a feature called *granulation*. Boiling material emerges from the surface below, pushing aside those cells already present. Here they release energy, mainly by releasing protons, whence their temperature reduces and they contract and sink back down again. They also shed material from around the cell boundaries and some from the cell surface, which contributes to the solar wind. The temperature of the returning surface material is thought to drop to around 4000K at the cell bases, at depths somewhere around 500 kilometres, where it is reheated. These granules of hot gas swell to between 1000 kilometres to 2000 kilometres across on reaching the surface and dwell there between 5 and 20 minutes, but more usually for about 10 minutes at a time.

Astronomers are able to observe sunspots on the far side of the Sun using a technique called *helioseismic holography,* and are thereby able to predict the appearance of solar activity, which may ultimately have an influence on the Earth. This technique again makes use of the acoustic pressure waves created by granulation, which have periodicities related to cell life cycles, and are in the region of 5 minutes or 3 millihertz.

Granulation is shown on the following page in the NASA modified picture originally taken in July 1997 by G. Scharmer of the Swedish Vacuum Telescope. Superimposed on it is a map of the USA, Canada, Central America and Greenland to provide an idea of scale.

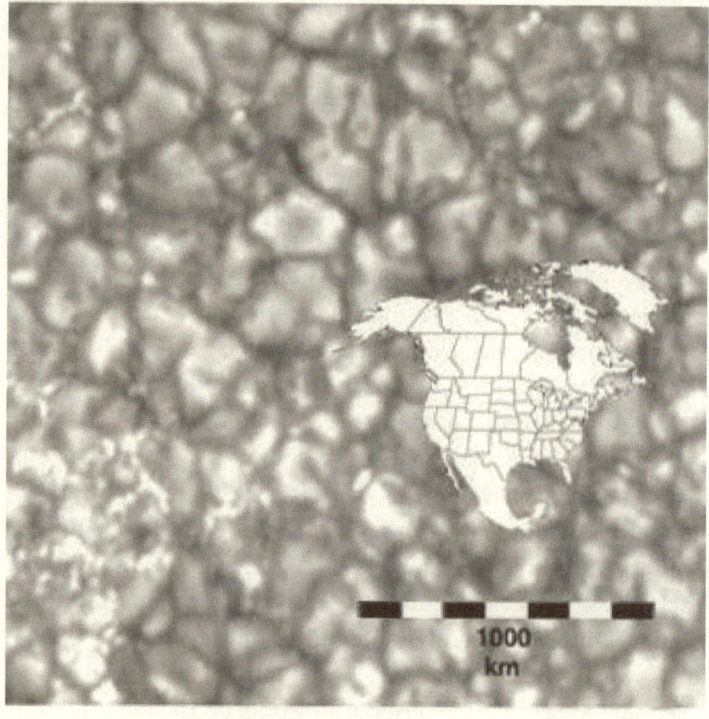

Figure 10   GRANULATION

The heart of the Sun is the *core*, the nuclear furnace that keeps the Sun alive by converting hydrogen into helium, and creating the other chemical elements known to exist. For this to occur, the core would need to generate temperatures of at least 15.7 million degrees, with a core pressure variously quoted as anything from 340 million to 225 billion times that of the Earth's atmospheric pressure. The diameter of the core is believed to occupy 20% to 25% that of the Sun. There are no definite interfaces or boundaries between the various regions and so these figures should be taken as a close guide only.

The *radiation zone* or *radiative zone* is the next region outside of the core. Quite simply, heat from the core radiates outwards towards the surface. Some textbooks offer a calculated estimate that it takes the plasma about 171,000 years to traverse the radiation zone. This zone is reckoned to occupy the region from the core to approximately 70% of the solar diameter. The radiation zone is cooler than the core with temperatures falling as the distance from the core increases. Average temperatures are estimated to be around 5 million degrees, falling to some 2 million degrees at its outer limits.

The third region is the *convection zone*, so named because it arises from the fact that from this point onwards, it becomes more efficient for heat energy to be transmitted by convection. Massive circulating cells of hot gas transport the heat to the base of the photosphere where the photospheric granulation cells, are born. These convection cells release energy into the photosphere, cool, and then sink back down again in a continuous cycle. The interface where the radiation zone ends and the convection zone starts is again fairly nebulous and is not defined.

When the convection zone "cells" reach the photosphere and granulation occurs, energy is released at the surface in the form of plasma, heat, light, radio waves, x-rays and gamma rays. Some of the charged particles will be ejected with enough force to reach escape velocity and fly off into space, to become the solar wind. Some material having attained lower velocities will fail to escape the Sun's gravity and will fall back onto the photosphere, as a kind of "rain".

There is a form of rudimentary atmosphere around the Sun, a thin layer called the chromosphere. The chromosphere is composed of most, if not all, of the elements present in the Sun as ionised gas. The main constituent by mass, accounting for about 70% of this very thin atmosphere, is hydrogen gas and about 28% is helium. Under normal circumstances, the chromosphere cannot be seen except during an eclipse because its detail is normally blanked out by the glare of sunlight. Nowadays, astronomers view the chromosphere with special telescopes fitted with *occulter discs*, which block out the light from the Sun's disc to give the same effect as an eclipse. In addition, some amateur telescopes are fitted with special filters called *calcium-k* filters that can look at the distribution of ionised calcium in the chromosphere, and from this, gain an appreciation of the structure of chromospheric magnetic fields.

Some satellites are monitoring the chromosphere looking at ionised iron and nickel spectral emissions. How this is achieved is detailed in the chapter *"What is the Sun Made Of?"* As already stated, the average temperature of the photosphere is 5776K, but just above the surface where the chromosphere begins, the temperature drops to around 4100°C or 4400K. From observational measurement, it is estimated that the chromosphere is between 2,500 kilometres to 10,000 kilometres thick, and at the higher altitudes, its temperature rises to between 30,000K and 100,000K. The increasing rise in temperature of the chromosphere with the rise in altitude is attributed to heating from the next outer region of the Sun called the corona, where temperatures between 1 million and 2 million degrees exist, with some sources claiming as high as 10 million degrees. Nobody knows why the coronal temperature rises to such levels although there are several theories.

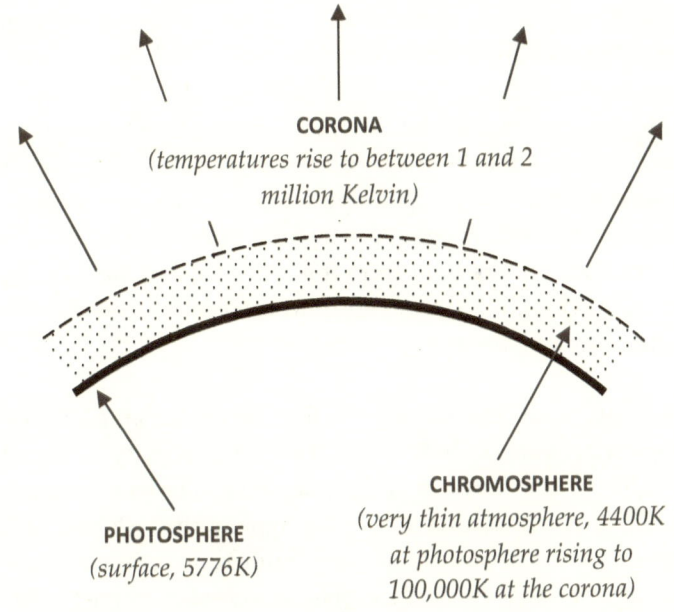

Figure 11   DIAGRAM OF THE SUN'S ATMOSPHERIC LAYERS

The corona is a vast region of rarefied hot plasma and ionised gas stretching out into space for several millions of kilometres. Although it is a million degrees or more, an astronaut would freeze to death in the corona. This is because the "temperature" energy originates from high velocity particles, such as electrons, which travel through the corona at around a quarter of the speed of light. The atmospheric "pressure" is extremely low, estimated to be at maximum about one ten billionth that of the Earth's atmosphere. The inside of an atom is predominately empty space with its constituent subatomic particles only accounting for a very small percentage of its overall volume. It is highly likely that a coronal high energy subatomic particle would pass clean through an astronaut without making any contact with any part of any atom in his body. Even an impact from a single electron will do no harm.

Under normal circumstances, the corona cannot be seen in visible light either. The corona is usually observed in ultraviolet light by using a special solar telescope, called a *coronagraph,* which is again a telescope fitted with an occulter disc to block out the light from the main body of the Sun, but since ultraviolet light is beyond the visible spectrum it is not diluted by daylight.

One feature of the corona is that it takes on a "spiky" appearance during periods of low solar activity. This is thought to be due to the emanating charged particles following the lines of the Sun's magnetic flux. When solar activity is high and there are numerous eruptions and emanations from the surface, the native solar magnetic field pattern becomes disrupted and distorted, and the observed coronal pattern becomes more random.

The nature of the photosphere, chromosphere and corona has been deduced from observation, whereas the internal structure of the Sun remains hidden from us. The model of the Sun's internal regions is a guess, but a guess based on helioseismic measurements, knowledge of its chemical composition, temperature, mass and average density, flavoured with an Earth founded knowledge of nucleosynthesis, plus some supposition regarding the convection zone. Details of how the various temperatures mentioned in this chapter are arrived at are given in the chapter *"How Hot is the Sun?"*

The Solar Wind

We have already seen that the Sun continuously discharges heat, light and material in all directions, in particular a hot plasma of charged particles, electrons and protons.

Granulation cells erupting onto the photosphere surface set up shock waves and eject this material into the chromosphere. In general, through astronomical observations, they are seen as jets or columns of material originating from around the cell boundaries in regions of high magnetic flux. These eruptions are called spicules, and can be likened to miniature flares. Spicules can last for about 5 to 10 minutes reaching altitudes between 3,000 to 10,000 kilometres. At any time, there can be around 100,000 spicules on the Sun's surface. Both flares and spicules are discussed in the chapter on *"Surface Features"*. As mentioned, the chromosphere's thickness ranges between 2,500 to 10,000 kilometres and so it is easy to see how this ejected material will attain escape velocities sufficient to propel it into the corona and onwards out into space. In addition to flares, there are occasionally more violent explosions on the Sun's surface, also discussed in the chapter on *"Surface Features"*, which can eject a vast amount of material in a single event.

The Sun is continuously emitting on average about *6.7 billion tons of material per hour,* or nearly 2 million tons a second! Only a portion of this material will be directed towards Earth, with the remainder flying off into space. The solar wind's speed is not constant but varies in general between 200 kilometres per second to 900 kilometres per second, or around half a million miles per hour to 2 million miles per hour. In round figures, it can take between 2 and 8 days to reach us, giving us time to take precautions when needed. The average wind velocity is quoted as 468 kilometres a second, a little over a million miles per hour. The average proton count in the solar wind is 8.7 per cubic centimetre, or 8.7 million per cubic metre. The electron and proton counts are usually quite similar. The solar wind high velocity particle flux densities should not be confused with the relatively few hydrogen atoms that inhabit interstellar space.

There are satellites continually monitoring this solar wind. Details of the proton counts per cubic centimetre, electron density, mean solar wind velocity and alpha particle (helium nucleus) density, are all published on the Internet and updated at regular intervals. One such satellite, at the time of writing, is ACE, which stands for Advance Composition Explorer. Launched by NASA in August 1997, it is now controlled by the US National Oceanic and Atmospheric Association, (NOAA), Space Weather Prediction Center, (SWPC). The on-board monitoring equipment is called SWEPAM, an acronym for Space Weather Electron Proton Alpha Monitor. Proton and electron density, solar wind speed, and the alpha particle to proton count ratio are updated at 10 minute intervals, and these and the other parameters mentioned here, can be seen on the *"www.spaceweather.com"* website, covering the last 24 hours up to the present time.

There is a *Planetary K Index*, or simply the Kp index, devised by Cornell University, USA, which maps the distribution of high energy particles around our polar regions and beyond, to forecast the extent of auroral activity. Other influences are monitored such as solar induced disturbances in the Earth's *magnetosphere*, (the magnetic field surrounding the Earth), the *Interplanetary Magnetic Field*, IMF, and the strength of radio emissions from the Sun on the 10.7 centimetre wavelength measured in *Solar Flux Units*, Sfu.

Historically, the Russian satellite Luna 1 first measured the solar wind in January 1959. The Americans followed this in 1962 with Mariner 2. Launched from the space shuttle "Discovery" on 6 October 1990, the Ulysses satellite has been monitoring the solar wind emitted from the Sun's Polar Regions. Just like the Earth, the Sun has a magnetic field with a North and South Pole. It was discovered that under the influence of the Sun's lines of magnetic flux, which are more concentrated at the poles, and which emanate directly outwards from its surface, the solar wind has a higher velocity than that emitted from around the Sun's equatorial regions.

However, it is only after the journey from the Sun is nearly completed, that satellites in near-Earth orbit provide us with accurate data, affording us limited time to react. They detect the solar wind around half an hour to an hour before it reaches the Earth. It is essential therefore, that astronomers and satellites continuously monitor the Sun's surface for visible signs of solar storms. In Brussels, the Solar Influences Data Analysis Center (SIDC) is home to the Regional Warning Centre (RWC). The SIDC collects and processes observations worldwide and then the RWC issues solar storm warnings as appropriate.

These charged particles of high energy electrons and protons, or plasma, also collect in bands encircling the Earth under the influence of our own magnetic field. First discovered in 1958 by the US satellite Explorer 1, they were named the *Van Allen Belts* after the designer of the equipment that discovered them. Electrons and protons tend to accumulate in the outer belt at distances around 15,000 kilometres to 60,000 kilometres,

(10,000 miles to 37,000 miles), above the Earth's equatorial regions, whereas the higher energy (heavier) alpha particles, and some protons, penetrate deeper into the inner Van Allen belt at altitudes up to 10,000 kilometres, or around 6,000 miles high. When solar activity intensifies, these Van Allen belts tend to "fill up" and excess plasma from the solar wind then spills over into our upper atmosphere around our North and South Poles, attracted by the flux concentration of the Earth's polar magnetic field. It is this, which is responsible for the creation of the auroras around our polar regions. The auroras occur high up in our atmosphere at altitudes from around 80 kilometres to 300 kilometres or from 50 miles to nearly 200 miles. At the highest altitudes, there is more oxygen than nitrogen and the solar wind electrons excite oxygen atoms to produce a rare red light, but at lower altitudes, from 300 kilometres down to 100 kilometres, oxygen gives off a green light. This difference is due to higher energy electrons penetrating further into the atmosphere, raising the excitation levels of the oxygen atoms where they emit light of a different colour. Nitrogen atoms are found in greater abundance at the lower altitudes and emit a blue/violet light, the violet being a mixture of red and blue. These processes are very similar to those used to create the orange glow from a neon light, and auroral conditions have been reproduced in the laboratory. Auroras have been seen on other planets in our solar system, Jupiter and Saturn, and possibly on Io, a moon of Jupiter, assisted by the magnetic influence of the mother planet. Many moons have no atmosphere and so cannot create aurorae but Io has a thin atmosphere of sulphur dioxide.

The influence of the solar wind distorts the Earth's magnetic field as shown in the diagram on the next page. The lines of flux in the field facing the Sun are compressed by the solar wind and those opposite are distended away from the Earth.

Under quiet conditions with a constant solar wind, at any given location on the Earth's surface, as the Earth rotates between day and night, that location will experience the transition from the compressed to the stretched magnetic field state. This effect is measured using an instrument called a magnetometer. Such a transition would be gentle, changing as the Earth revolves. However, this rate of change can be abrupt with the onset of a solar storm, and can peak in as little as half an hour.

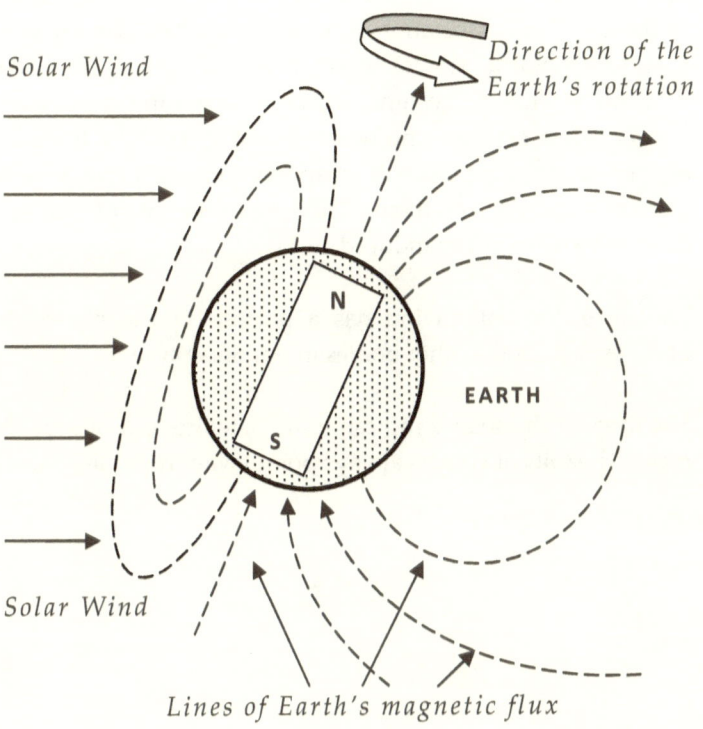

Figure 12 DISTORTION OF THE EARTH'S MAGNETOSPHERE BY THE SOLAR WIND

It is the changing magnetic flux that induces heavy currents into long metal conductors such as pipelines, railway lines or power cables. Violent solar storms can practically strip away the Earth's magnetosphere altogether, effectively opening a portal to the incoming solar wind virtually to the Earth's surface.

If a current is passed down an electric wire, that wire becomes surrounded by a magnetic field. Michael Faraday first noticed this in the early 19th century, when switching on and off an electric circuit caused the needle in a nearby magnetic compass to flicker. This association between an electric current and a magnetic field is similar to that which makes an electric dynamo work. A dynamo works by spinning a bundle of copper wires between the poles of a magnet. The bundle of copper wires experience a changing magnetic field which induces a current into them. The greater the rate of change of field, the greater is the induced current.

Therefore, the solar wind has a profound influence on the Earth as well as the other bodies in our Solar system.

The force of the solar wind can move satellites out of orbit and makes the tails of comets always point away from the Sun!

How Hot is the Sun?

We know that the Sun must be hot from the simple fact that on sunny days it is warmer than on cloudy days. From classical physics we know that there is a relationship between the temperature of a hot object and the colour of the glow it emits.

If you heat up a bar of metal, iron for example, to start with it will appear unchanged. Heat it further and a dull red glow will appear. Heat it further still, and the red glow will brighten, then turn orange, and yellow until it eventually melts. If you burn magnesium, as used in signal flares, it burns with a very hot bright bluish white flame.

Clearly then, there is a relationship between the temperature of something and the colour it glows. However, the same is not true for a lump of wood. To start with there will be no colour change. At higher temperatures, it will char and blacken before eventually igniting. The situation here is quite different. The wood is undergoing chemical change during the heating process, carbonising and releasing gaseous compounds, whereas the iron bar in the text above retains its chemical identity.

Physicists have performed many experiments to determine how different types of surface absorb or radiate heat energy. Dull black surfaces are good at both absorbing and radiating heat energy whereas bright shiny surfaces are good at reflecting heat and light. During the course of their experiments, scientists devised a piece of apparatus called a *black body radiator*. It is an enclosed chamber with blackened inside surfaces and a single small hole providing visual access to its interior. The concept is that an ideal black body will absorb all incident electromagnetic radiation, such as heat and light, and, unlike a mirror, will not reflect or retransmit any of it. It follows that such a device is an equally efficient incandescent or thermal *radiator*.

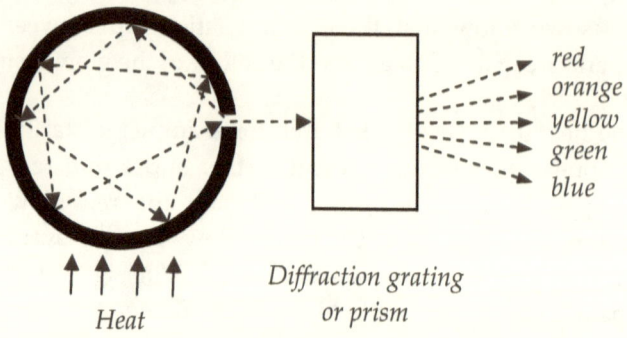

Heat

Diffraction grating or prism

red
orange
yellow
green
blue

Figure 13   THE PRINCIPLE OF BLACK BODY RADIATION

The schematic diagram above illustrates how a heated black body radiator releases energy in the visible light, infrared (IR) and ultraviolet (UV) regions and radiates them within its interior. In the laboratory, its emissions would then be disseminated into a continuous spectrum for analysis.

Just like the piece of iron, when the black body radiator is heated, the inside will eventually start to glow. As the temperature rises further, the intensity of the glow will increase, and, again like the piece of iron, the colour of the emitted light will change. Different colours can be created by the combination of two or more other colours. Yellow, for example, may be created by combining red and green light. Pure white light is a mixture of all the colours. Whereas incandescent light across the visible spectrum will appear at elevated temperatures, invisible light in the infrared and ultraviolet regions will be present at lower and higher temperatures respectively.

Light from the black body radiator is converted into a continuous spectrum and the intensity of the various spectral components is measured. When combined, the two sets of measurements produce a curve, which becomes the signature of the spectral emissions for a particular temperature. This is illustrated schematically in the following graph.

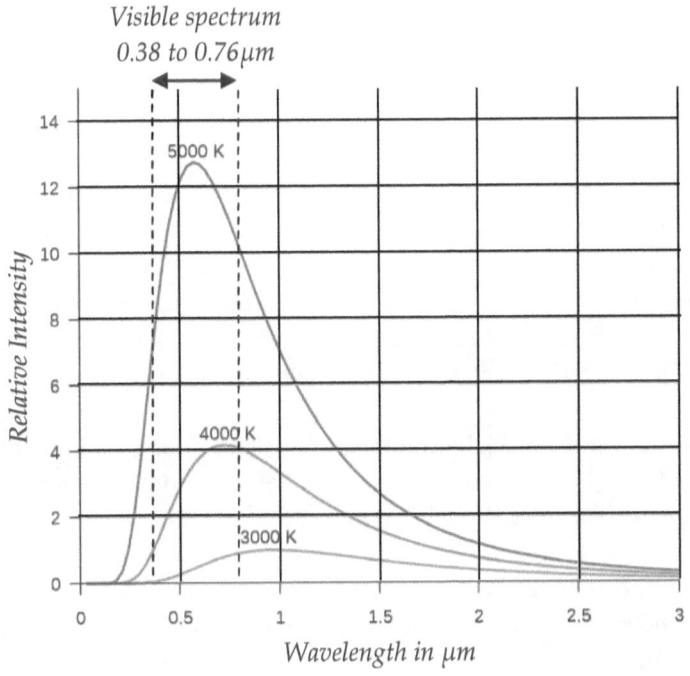

Figure 14   GRAPH SHOWING THE RELATIONSHIP BETWEEN BLACK BODY RADIATION AND TEMPERATURE

The unit of measurement (μm) under the *Wavelength* axis is *microns*, or one millionth of a metre. The visible spectrum ranges from around 0.38 μm at the blue/violet end, to around 0.76 μm at the red end. Shorter wavelengths venture into the ultraviolet and thence the extreme ultraviolet (EUV) regions. Likewise, the longer wavelengths span the infrared territory.

Temperatures are quoted in Kelvin, designated by the letter "K", where 273.16 Kelvin equals 0° Centigrade. The graph illustrates how the intensity increases as the temperature rises, but it also shows how the peaks of the curves migrate left towards the shorter (blue) wavelengths. This demonstrates that as the temperature rises there is a colour shift, initially forsaking red components, then turning increasingly blue. Therefore, there is a discrete composite colour/intensity signature for any prevailing temperature.

The Sun is, in practice, a black body radiator and by carefully measuring its surface colour components with their attendant intensities, produces a discrete curve which reveals that the average surface temperature of the Sun's photosphere is 5,776K, or 5,503°C. There are some localised hotspots reaching 6,400K, together with sunspots, which are cooler.

The use of the black body radiation graph can be simplified by applying *Wien's Law*. Published in 1893 and named after the German scientist Wilhelm Wien, (1864 to 1928), it states that the underlying shape of all the curves are essentially the same, and one only need consider the wavelength, λ, at the very peak of the curve in order to determine the temperature, using the formula:---

$$\lambda = \frac{0.0029}{T}$$

where the wavelength is in metres and the temperature T is in Kelvin. The formula can be transposed thus:---

$$T = \frac{0.0029}{\lambda}$$

Given that the peak wavelength of the Sun's emission curve occurs *approximately* at 0.5 μm, (5 x 10^-7 metres), which is actually in the *green* section of the visible spectrum, it provides an approximate average photospheric temperature of 5,800K.

Similarly, using this technique, the temperature of other surface features such as sunspots can be determined. Sunspots look black on initial inspection only because they are cooler than the surrounding photospheric temperature. Taken on their own merits they do have their own colours, which indicates that they generally lie in the range 3,000K to 4,500K, or 2,727°C to 4,227°C. The "colour temperature" of the chromosphere measures around 4,500K just above the photosphere rising to around 20,000K as it approaches the corona. Textbooks give various values due to the fact that the interface between the chromosphere and corona is indistinct.

However, the Earth's atmosphere with its attendant pollution can distort the Sun's emission spectra and give a false result, so satellites take modern measurements exo-atmospherically.

Another method of determining the Sun's surface temperature is by using the *solar constant*. The solar constant is a measure of the solar flux energy arriving at the Earth. The value currently assigned to this flux is *on average* 1,366 Watts per metre squared. Using an instrument called a pyrheliometer; (see *Glossary*), this measurement is also generally made exo-atmospherically, due to absorption and pollution. Early measurements were made using high altitude balloons. Values taken in Germany at ground level have yielded figures between 800 and 1,100 Watts/metre². The constant varies with the time of year, fluctuating by 6.9%, reaching 1,412 Watts/metre² in January when the Earth is at that point in its orbit when it is nearest to the Sun, (called *perihelion*). In July, it drops to 1,321 Watts/metre², when the Earth is at its furthest point in its orbit from the Sun, (called *aphelion*). However, we can also use the solar constant to calculate the total power that the Sun radiates per second, called its *luminosity*. Since we receive 1,366 Watts/metre² at an average distance of 150 million kilometres, every square metre of an imaginary sphere surrounding the Sun, having a radius of 150 million kilometres, (radius, $r = 150 \times 10^9$ metres), will also receive that level of flux. The surface area of a sphere is $4\pi r^2$ so the luminosity, L, will be:---

$$L = 4\pi r^2 \times \text{solar constant} = 3.86 \times 10^{26} \text{ Watts,}$$

which is 386 billion billion megawatts.

If we now refer the luminosity back to the actual solar surface we can calculate the flux density at the photosphere. Now the sphere becomes that of the Sun itself. Its diameter is 1.392 million <u>kilometres</u>, so its radius, "rs", is 696 million <u>metres</u>. The Sun's radiated energy flux will be:---

$$\frac{L}{4\pi rs^2} = \frac{3.86 \times 10^{26}}{4\pi \times 696^2 \times 10^{12}}$$

which is 63.41 megawatts per square metre.

Again we can use these values to check the Sun's average surface temperature, and validate the decision to consider the Sun as a black body radiator by using *Stefan Boltzmann's Law*.

Ludwig Boltzmann (1844 to 1906) and Jozef Stefan (1835 to 1893) were two Austrian physicists who together investigated the properties of black body radiation. Their law stated that the energy flux emanating from a surface, in Joules per second per square metre (or Watts per square metre), was proportional to the fourth power of the temperature, in Kelvin:---

Flux = σT^4,   where σ is called *Stefan's constant.*

The constant σ was determined experimentally to be:---

$$\sigma = 5.67 \times 10^{-8} \text{ Watts/metre}^2/K^4$$

Note: The unit of energy, the "Joule" is defined on page 113.

Therefore, simply transposing the formula gives the Sun's average photospheric temperature as the fourth root of the surface energy flux, divided by Stefan's constant:---

$$T = \sqrt[4]{\frac{63.41 \times 10^6}{5.67 \times 10^{-8}}}$$

$$= \underline{5,782K}$$

Clearly, as this answer differs by as little as 6 degrees, the assumptions made are justified.

It is an interesting fact to note that by turning these calculations around, it can be shown that the amount of energy arriving from the Sun would only heat the Earth to an average annual temperature of 209K or -64°C. The temperatures that we actually experience are due to the additional heating of the surface from the Earth's molten core.

It has been stated that the Sun is composed of gases and plasmas. Plasma is a gas, which has been heated to such a degree that electrons are stripped off the various atoms, and since they are at such a high level of excitation, recombination becomes impossible. This means that the plasma will contain numbers of free electrons and due to the ionisation of hydrogen, free protons. As such, plasmas are electrically conductive and are influenced by magnetic fields. During the collapse of a plasma following the removal or reduction of the ionising temperature, the electron – ion recombination process will release electromagnetic energy. As a gas is heated, the atoms become agitated and eventually start to collide with each other. The temperatures involved result from high velocity collisions between atoms.

Measuring the temperature of plasma is not a straightforward affair. The principle can be demonstrated in the following simplified example. A molecule, atom or electron in motion will have a kinetic energy equal to $\frac{1}{2}mv^2$ where "m" is its mass and "v" its velocity. This kinetic energy can be equated to temperature using a physical constant known as *Boltzmann's constant*, which is generally identified by the letter "k". The relationship is:--

$$\tfrac{1}{2}mv^2 = {}^3\!/_2 kT, \qquad \text{(where T is in Kelvin)}$$

Therefore an electron, with a mass of 9.11×10^{-31} kilogrammes, travelling with a velocity of 6,700 kilometres per second, has an energy (temperature) equivalent to 1 million Kelvin.

Note: The Sun's corona is estimated to have temperatures of a million degrees or more due to the very high velocity of atoms and particles present in what is a very rarefied gaseous region. In actuality, comets, which are effectively dirty snowballs composed of rock, dust and ice, have been seen to survive for significantly long periods when entering the Sun's corona. This is due to the fact that the internal region of an atom is mostly empty space, and a high velocity/high temperature electron or proton can easily pass straight through it without making any contact. The *mean free path* of a coronal electron, or the average distance it might possibly travel before colliding with something, has been calculated to be several kilometres.

Plasmas start to form at temperatures around 5,000°C and commercial plasma metal cutters operate at temperatures around 25,000°C.

To determine the temperatures of astrophysical plasmas, use is made, amongst others, of equations called the *Saha Ionisation equation*, or the *Saha Langmuir equation*. Developed from a combination of classical physics and quantum mechanics, these equations are considered to be beyond the scope of this book. Broadly, these equations look at the ionisation energies of a gas that can then be related to electromagnetic emissions and temperature.

Flares are ejections of plasma and emissions of electromagnetic radiation covering a wide spectrum from radio waves to x-ray bands. As such, they have external temperatures of a few million degrees but can have internal temperatures anywhere between 10 and 20 million degrees. Different techniques are used to measure these temperatures, which are performed by measuring their x-ray signatures. X-rays cannot penetrate Earth's atmosphere without severe attenuation and, again, satellites operating outside the Earth's atmosphere perform all such measurements.

How We View the Sun?

Before we start exploring various surface features on the Sun, such as sunspots, and try to unravel their mysteries, it is worth exploring exactly how the Sun's cardinal points are designated. In addition, we will need to explore a subject called *"Heliodetics"*. It is a technique, which enables us to determine the *heliographic* coordinates of some feature or other, on the Sun's surface, the solar equivalents of latitude and longitude used here on Earth, and is the solar equivalent of the Earth science subject *Geodetics*.

When we look down on an atlas or a map, the convention is that we view it with the North uppermost, South below, and the West to our left hand side and the East to our right. The same would apply to a man in space, possibly on the surface of the Moon, looking down at the Earth.

If a man were standing on the North Pole looking out into space, towards the Sun, he will have actually performed an "about face." Now, his left and right-hand configurations will have become reversed. In effect, looking toward the Sun from the Earth he would be viewing the Sun from behind. Both the Sun and the Earth's West are now on his right hand side, with the East on his left. Therefore, the Sun's East and West appear transposed. The Sun's East is now on his left hand side, and not the West as previously. This is shown in the diagram on the following page. Although we have considered how a man situated at the North Pole might see things, the same orientation holds true for an observer at any latitude in the Northern hemisphere.

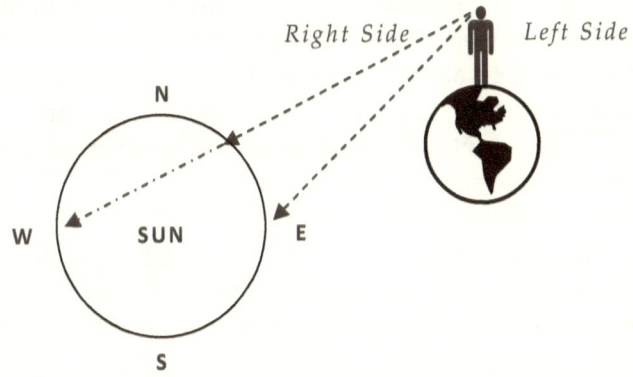

Figure 15   REVERSE ORIENTATION OF THE SUN AS SEEN FROM THE EARTH
(THE OBSERVER IS LOOKING AT THE BACK SIDE OF THE SUN)

In our Southern hemisphere, as shown in the diagram below, the situation is reversed and a man looking up at the Sun will view it inverted with its South Pole uppermost.

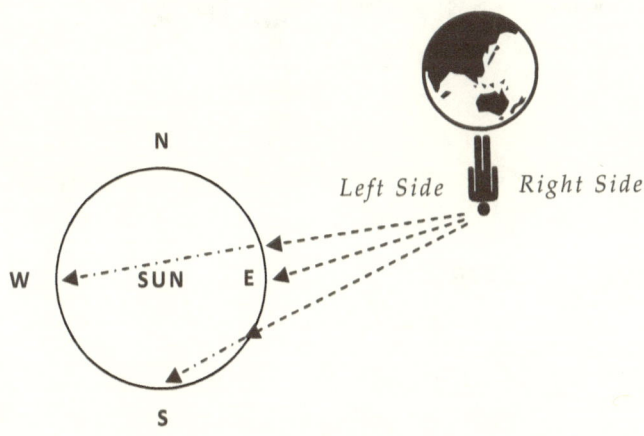

Figure 16   THE SUN'S ORIENTATION APPEARS INVERTED BETWEEN
EARTH'S NORTHERN AND SOUTHERN HEMISPHERES

49

His left and right hand sides are also transposed relative to the man in the Northern hemisphere and so he will view the Sun's East on his right and the West on his left. Once again, this orientation will be the same for an observer at any latitude in the Southern hemisphere and only changes to the previous (northern) orientation as one crosses the equator.

This East–West dichotomy is the explanation as to why published images of the Sun are presented either with the conventional cardinal point orientation, or with the "East to the left" variant.

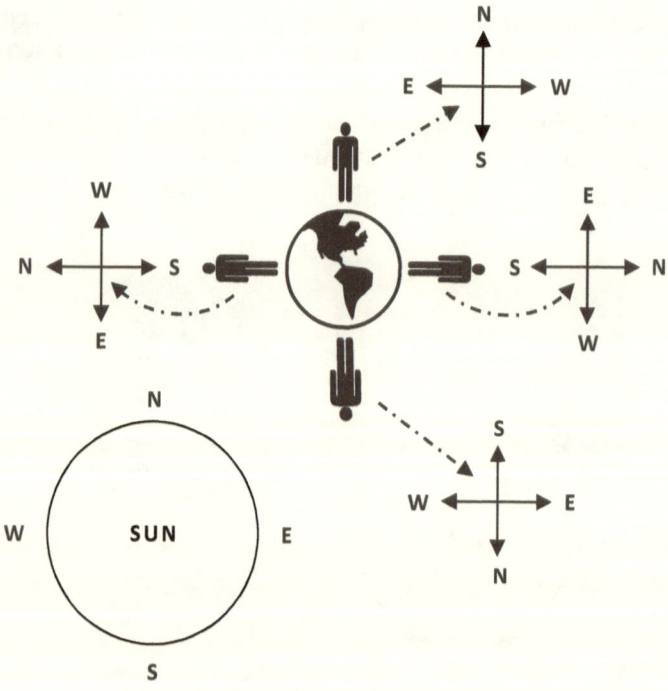

Figure 17   COMPLETE DIAGRAM SHOWING SOLAR ORIENTATIONS
AS VIEWED FROM THE EARTH'S CARDINAL POINTS

For the sake of completeness, the previous diagram shows the whole picture, including the orientation of observers standing on the Earth's equator. As the Earth rotates, the Sun rises in the East, travels across the sky during daylight hours and sets in the West. This will take around half a day, with the remaining half of the Earth's daily revolution taking place during the night. Therefore, an observer standing on the equator looking East will see the Sun rise, and as it travels overhead its orientation will rotate in a clockwise manner. It will ideally change from its Eastern orientation at sunrise to its Western orientation at sunset, twelve hours later, continuing to rotate in a clockwise manner around the back of the Earth returning to its original position, at the following sunrise.

This second order change of orientation will manifest itself at all other latitudes, and cants the observed North Pole axis of the Sun through its daily excursion across the sky. This angular change of the Sun's pole of rotation over time is termed the *parallactic angle* or *angle of parallax*. The angular change has maximum effect at the equator where the Sun's cardinal points simply appear to rotate during the course of the day. However, there is no skewing of the observed alignments at the Earth's poles.

The term *solar noon* is given to that time when the Sun has reached its highest altitude at the particular observer's site or meridian. Although solar noon can occur at 12:00:00 UT under certain circumstances, say at Greenwich, England, it may well occur at quite a different time dependent both on the observer's location and the particular day of the year.

A standard of time universally applied to any location on the Earth is *Greenwich Mean Time*, GMT, or *Universal time*, UT. Sometimes used is the acronym UTC standing for *Coordinated*

Universal Time that is derived from an atomic decay clock and not based on solar mechanics. However, GMT, UT and UTC all relate to the same *synodic* 24 hour *mean solar day*.

As an example, consider an observer in the Northern hemisphere at latitude 50°. At solar noon, the Sun's North Pole will be uppermost, but at sunrise and sunset, it will be canted over by (90° - 50°) = 40°, as shown in the diagram below.

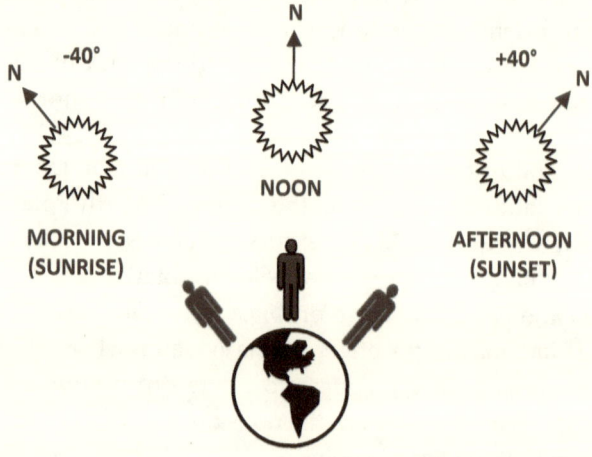

Figure 18 SIMPLIFIED DIAGRAM OF THE CHANGING ATTITUDE OF THE SUN'S DISC THROUGHOUT THE DAY FOR AN OBSERVER AT LATITUDE 50⁰ NORTH

At dawn, the Sun rises in the East and so the Sun's attitude will become modified by an element of that orientation pertinent to the Earth's Eastern equator. That is, it will be rotated anticlockwise by 40°. At sundown, the Sun sets in the West and so its polar attitude will now be modified by a 40° clockwise component of the Sun's Western equatorial configuration. Therefore, during the course of the day, the attitude of the Sun's North Pole swings clockwise from a -40° orientation at sunrise, through True North at the solar noon, to the +40° orientation at sunset.

The Sun only rises precisely in the East, and sets precisely in the West at the *Equinoxes*. During the summer months, the Sun ascends higher in the sky compared to the winter months. Sunset occurs later and the hours of daylight are longer. Twice a year a situation arises whereby the length of daylight nearly equals the length of night at 12 hours each. The time of "equal day" and "equal night" is called an equinox. The *vernal equinox* occurs on either March 20 or March 21 annually depending on whether or not it is a Leap Year. The *autumn equinox* falls on either September 22 or September 23.

It is these differences in the length of the day and the maximum height reached by the Sun at solar noon, at the height of summer and the depth of winter, which reveals the fact that the Earth's Pole of Rotation is not vertical.

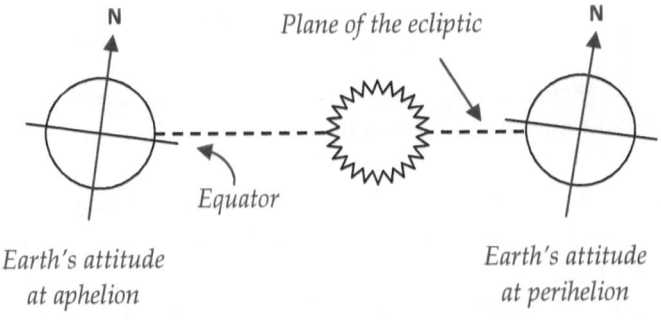

Earth's attitude
at aphelion

Earth's attitude
at perihelion

Figure 19   THE EARTH'S POLE OF ROTATION IS CANTED RELATIVE TO THE PLANE OF THE ECLIPTIC

The Earth's rotation axis is canted over by 23½° relative to its plane of orbit around the Sun, the plane of the ecliptic, as shown in the figure 19.

Consider the situation in the Earth's Northern hemisphere. During the summer, the Earth passes through aphelion. The Earth's pole of rotation tilts *generally toward* the Sun and so the Sun appears to ascend higher in the sky. The reverse occurs during the winter. Whilst the Sun passes through perihelion, the Earth's pole of rotation tilts *roughly away* from the Sun and so it appears lower in the sky. At the equinoxes, the Earth's pole of rotation will be vertical and the Earth's equator will be aligned to the plane of the ecliptic.

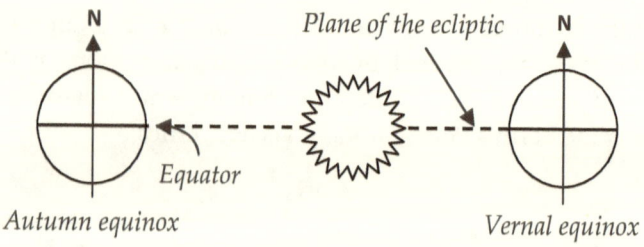

Figure 20   AT THE EQUINOXES, THE EARTH'S POLE OF ROTATION IS UPRIGHT AND THE EQUATOR ALIGNS TO THE PLANE OF THE ECLIPTIC

The maximum deviation of the pole of rotation angle, (called *declination*), subtended to the plane of the ecliptic, occurs at the solstices and is equal to the angle of the Earth's tilt. In the Northern hemisphere, the summer solstice occurs on June 22, and the Earth's pole of rotation tilts *directly toward* the Sun. In the Southern hemisphere, this would become the winter solstice.

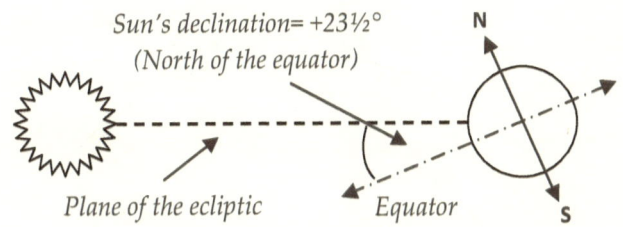

Figure 21   SUMMER SOLSTICE IN THE NORTHERN HEMISPHERE

On the summer solstice, the Sun reaches its highest altitude in the sky and we get the longest day. Although this is true for both Northern and Southern hemispheres, the dates of these events will be reversed, June 22 in the North, December 22 in the South.

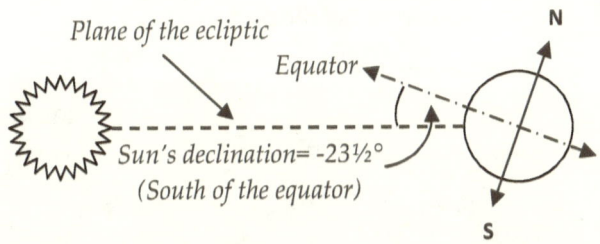

Figure 22   WINTER SOLSTICE IN THE NORTHERN HEMISPHERE

Winter solstice in the Northern hemisphere occurs on December 22 when the Sun reaches its lowest daytime altitude, and we get the shortest day of the year. Here the Earth's pole of rotation tilts *directly away* from the Sun.

Therefore, the difference in the Sun's altitude at the two solstices will be twice the tilt angle of the Earth's pole of rotation, 47°, shown in the next diagram. The different path lengths in figure 23 illustrate why the length of the days vary.

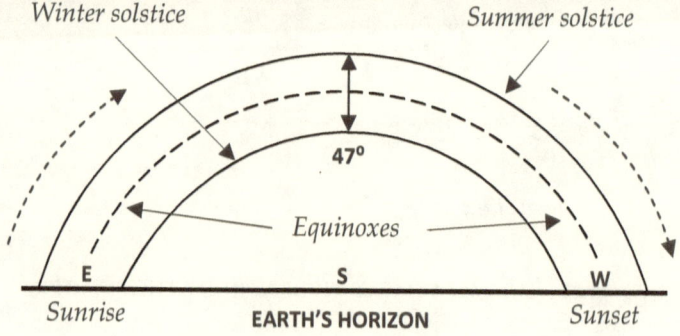

Figure 23   DAILY PATH OF THE SUN ACROSS THE HORIZON IN THE NORTHERN HEMISPHERE

From this it can be seen that during the Northern hemisphere's summer, the Sun rises North of East, and in the winter, South of East, and is shown in the diagram below.

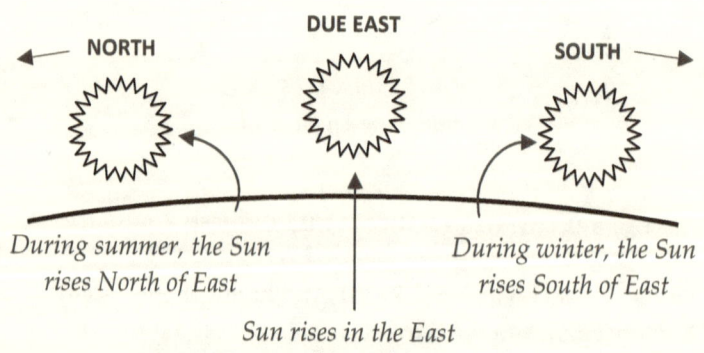

Figure 24   SUNRISE IN THE NORTHERN HEMISPHERE

Therefore, in the example illustrated in figure 18, the parallactic angle can exceed the stated 40° that occurs at the equinoxes. In fact, it will achieve angles up to 44.5°. In summer, the Sun rises earlier and sets later in the day accommodating a greater portion of the equatorial East to West angular transition. During the winter months, later sunrises mean that the range of parallactic angles occurring will be smaller.

The parallactic angle can be calculated for any time and any observer's latitude. There are freeware programs available on the Internet, which will perform these calculations once the observer has input the time and the coordinates of his or her location. Two programs available at the time of writing are the *"TiltingSun"* program and the *"Helio Viewer"* program. In addition to providing the parallactic angle, both these programs provide the user with a host of related parametric and heliographic data. The TiltingSun program, by Les Cowley, can be downloaded from the *"Atmospheric Optics"* website *www.atoptics.co.uk*. The Helio Viewer program is one of a cache of three related programs, all of which can be downloaded from the website *www.petermeadows.com*. (Peter Meadows is a member of the Solar Section of the British Astronomical Association).

Both these programs are well supported with User Guides and Help facilities.

The data provided by these programs will be essential in determining the heliographic coordinates of some surface feature of the Sun needed for recording or reporting purposes.

How Do We Know the Sun Rotates?

Simply by looking at the Sun, of course only through properly filtered apparatus and by no other means, one would not be able to tell that the Sun is actually rotating about its own axis. Further, one would be unable to discern that its pole of rotation is actually skewed by about 7¼° from the vertical, or relative to the plane of the ecliptic.

This means that the apparent line of its equator, that it presents to an observer on Earth, does not equally bisect the Northern and Southern hemispheres as seen, except for two days each year, one in July and the other in December. Further, taken in conjunction with the fact that the Earth's own polar axis tilts over by 23½°, the maximum observed swing of the Sun's polar axis, when seen from the Earth, swings from left to right or East to West, by ±26.28°. If it were possible to see this phenomenon, it would appear as if the Sun was wobbling from left to right throughout the course of the year. To be able to plot heliographic coordinates, these continuously changing conditions need to be determined and accounted for.

So, how do we know the Sun's polar axis is skewed? How do we determine exactly where the line of the Sun's equator actually lies on a nebulous ball of hot gas?

Well, it works like this!

These phenomena were discovered by studying the pattern of behaviour of many thousands of sunspots observed over a period of some four centuries. The first thing noticed was that sunspots tended to migrate across the face of the Sun from

East to West. Neither sunspots nor sunspot groups ever stayed in one place, nor did they travel in the opposite direction. This simple observation suggested that the Sun was rotating about a polar axis. The next thing that needed to be determined was the rate of rotation, but this did not turn out to be as easy as it was first thought. Two factors were in play. The first one was that, as mentioned, the Sun does not rotate vertically on its axis. The direction of travel of the sunspots was not absolutely horizontally aligned to what was originally thought to be the Sun's equator. The second fact discovered was that the rate of progress across the Sun's disc was quickest around the equatorial regions, but became progressively slower at higher latitudes both North and South.

We shall first consider the Sun's pole of rotation and the position of the solar equator.

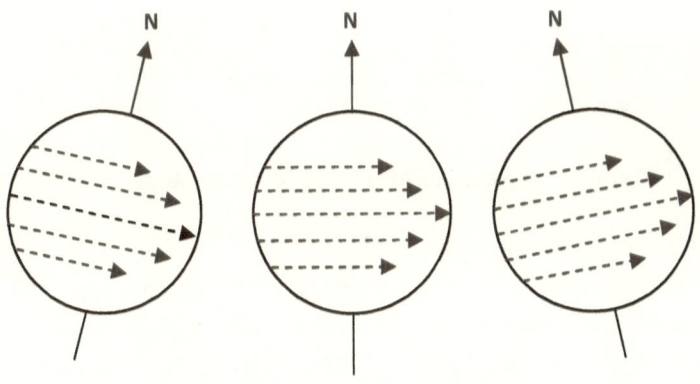

Figure 25 PROGRESSION OF SUNSPOTS ACROSS THE SOLAR DISC

The diagram above depicts the East to West progression of sunspots or sunspot groups across the face of the solar disc and the manner in which they align to the actual solar equator.

As previously stated and shown schematically, sunspots or sunspot groups on the equator will migrate faster than those at higher latitudes. The diagram has been simplified to show how the axis of the Sun's pole of rotation can be determined by tracking their progress. Since the Sun's solar axis will be tilted towards us at some point in time, sometimes upright, and away from us at other times, the line of the solar equator will displace northwards and southwards through an angle, which will describe the 7¼° tilt in the axis of rotation. This changing angle is denoted as B_0, (B nought). The diagram below shows the disposition of the solar equator throughout a terrestrial year.

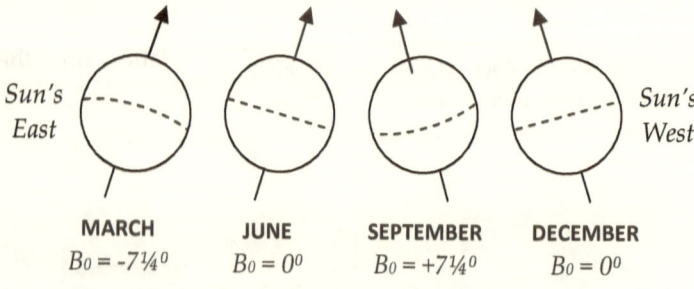

Figure 26   ANNUAL PERTURBATION OF THE SOLAR EQUATOR

The Sun tilts away from us in March when the displacement in the alignment of the Sun's equator, B_0, reaches the -7¼° maximum. The Sun's lines of latitude therefore are all displaced upwards by this angle. The opposite occurs in September when the Sun is tilted toward us. In June and December, the Sun's equator is normal to that of the Earth, and B_0 is zero. However, as can be seen, the Sun's polar axis still cants to the West in June and to the East in December. We have seen from the previous chapter that the Earth has a 23½° tilt in its pole of rotation relative to the plane of the ecliptic, which is responsible for our seasons. Now, these two axial tilts, taken in combination, result in the previously mentioned maximum observed axial swing of ±26.28°, shown in the diagram below. This pole of rotation angle is denoted as "P".

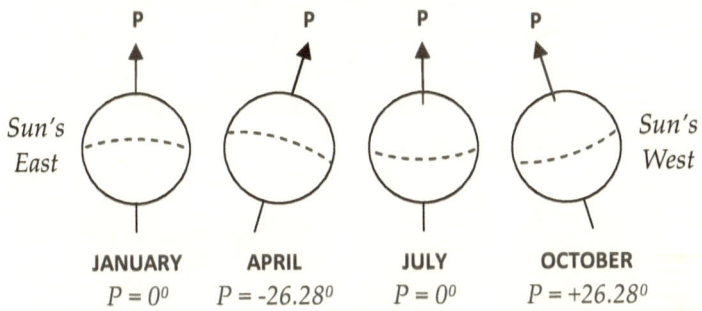

| JANUARY | APRIL | JULY | OCTOBER |
| $P = 0^0$ | $P = -26.28^0$ | $P = 0^0$ | $P = +26.28^0$ |

Figure 27   ANNUAL CYCLE OF THE SOLAR POLE OF ROTATION

The swing between maximum positive and maximum negative takes six months but the actual day on which this falls may vary due to the inclusion of Leap Years.

We shall now explore the variability in the migration rate of sunspots or sunspot groups at various latitudes. From numerous records, it was statistically determined that the rotation rate at the true solar equator (B_0) is 13.72° per hour. Refer to the chapters "*Sunspots*" and "*The Magnetic Sun*". This meant that the solar equator had a (synodic) rotation period of 26.24 days. Sunspots rarely occur above latitudes of ±40°, but by tabulating the transit rate of sunspots at various latitudes, it was established that the Sun has a differential rotation. Found by interpolation, the (synodic) polar rotation periods extend to around 36 days. Exact figures are not quoted because they cannot be confirmed by measurement.

We now need to understand two terms used in astronomy that define rotation periods, *sidereal* and *synodic*.

A sidereal rotation is the period taken for a rotating body to turn through 360°. A synodic period is the time taken for a body to rotate to a point where it presents the same face to a second body orbiting around it. In other words, it is the sidereal period, plus the extra time needed to catch up to the new position attained by the second orbiting body. This is shown in the diagram below.

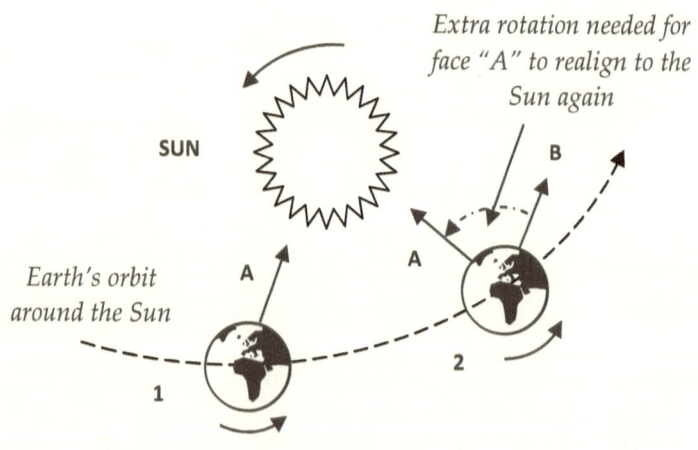

Figure 28   SIDEREAL AND SYNODIC ROTATIONS

The Earth, at position 1 presents face "A" towards the Sun. The Earth and the Sun both rotate anticlockwise as viewed from above, looking down onto their North Poles. After one sidereal day, where the Earth has rotated through 360°, face "A" now becomes face "B" shown at position 2. To present face "A" to the Sun again at position 2, to complete one synodic day, the Earth must rotate a little further, and take a little longer. Exactly the same situation exists for the Sun and it, too, will describe both sidereal and synodic conditions.

It is an easy matter to measure the duration of the Earth's sidereal day. All that is required is to accurately plot the position of a distant star, and then measure the time it takes to retake that position. A star can be chosen as a reference since the Earth does not orbit around it. The duration of a terrestrial sidereal day is 23 hours, 56 minutes and 4.2 seconds, and not 24 hours! The Earth makes one orbit of the Sun in approximately 365.25 days, accounting for Leap Years. Again, the exact annual duration can be determined by plotting the precise position of (many) stars and measuring the time taken to assume the same position the following year. The 24 hour day is the synodic day, also named the *"mean solar day"*. Therefore, in 365.25 days, the Earth completes a 360° orbit of the Sun, so a year takes (24 x 365.25) = 8766 hours. The sidereal day is 23.9345 hours, expressed as a decimal. So in 8766 hours, the Earth completes 8766 ÷ 23.9345 = 366.25 revolutions. To complete a synodic day the Earth will have to rotate an extra (365.25 ÷ 366.25) ° = 0.99727°, making a total of 360.99727°.

$$\text{(23 hours, 56 minutes, 4.2 seconds)} \times \frac{360.99727}{360} = 24 \text{ hours}$$

$$\underline{\text{Sidereal}} \qquad\qquad\qquad\qquad \underline{\text{Synodic}}$$

The synodic day is called the *mean* solar day because the Earth's orbit is not circular but slightly elliptical. As such, the actual duration of each day, one to another, will differ on average by about 11 seconds. This is dealt with in detail in the chapter *"The Equation of Time"*.

Now having determined the duration of the terrestrial sidereal and synodic periods, those of the Sun can be determined, and, once again, use is made of the migration patterns of sunspots. The Sun's disc is half a sphere with the equator representing a span of 180°. We can plot the hourly angular migration, both near the equator and at higher latitudes where it is known that the Sun's orbital rate is slower. We can now import into these measurements details of our own rotation and thus determine the true solar rotation rates for the various latitudes.

Earlier in this chapter, it was stated that the average solar equatorial synodic orbital period is 26.24 days. Synodic periods change daily due to the eccentricity of the Earth's orbital path, and so in general use tends to fall from grace in favour of non variant sidereal periods. The equatorial sidereal period is 24.47 days, and so to find the sidereal period at higher latitudes, an empirical formula has been devised which best matches the observed conditions. It is included here for those interested, where L is the latitude in degrees:---

$$= 14.713 - 2.396 \sin^2 L - 1.787 \sin^4 L \text{ degrees per day}$$

Note: "Sin" is an abbreviation of the trigonometrical function "sine", which is defined as the angle relating to the ratio of the the length of the opposite side of a right angled triangle divided by the length of the hypotenuse, (the longest side).

NASA uses a modified version of this formula, which has been standardised on a latitude of 16°:---

$$= 14.37 - 2.33 \sin^2 L - 1.56 \sin^4 L \text{ degrees per day}.$$

The 16° latitude was chosen due to some uncertainty in assessing sunspot activity precisely on the equator itself. These formulae hold good for latitudes up to around 50°. Due to the general absence of sunspot activity beyond these latitudes, the accuracy of calculations in those regions cannot be verified. As such, the situation at the poles is somewhat uncertain, but the current best guesses are that the sidereal period is 34.19 days and the average synodic is 36.7 days, though some sources quote 38 to 40 days.

In the chapter *"Carrington Rotations"*, an average synodic period of 27.2753 days has been adopted, being the period at Latitude 26°. This particular latitude was chosen because it is the latitude at which the greater part of sunspot activity occurs throughout the course of the solar cycle.

The characteristics of solar rotation have nearly all had to be determined by analysing countless records of sunspot existence. Following their conception, sunspot groups may well evolve during their journey across the Sun's disc, which only serve to confound analyses. Single sunspots generally provide a more reliable and accurate picture. Therefore, the determination of solar differential rotation parameters cannot be an exact science, but given the quantity of sunspot and group records now available a statistically significant precision can be awarded to this data.

In recent times, scientists have undertaken helioseismic investigations of the Sun and are developing an understanding of the motion of the inner regions.

The Prime Meridian

Before we investigate how to navigate around the nebulous surface of the Sun, it is worth reviewing how it is achieved on Earth. Unlike the Earth, which has numerous fixed and unchanging geographical features, the surface of the Sun is, at least in part, featureless except for the random and transient appearances of individual sunspots and sunspot groups.

Through the endeavours of ancient explorers, maps were drawn which eventually encompassed our entire globe. Maps identify geographical features and spatially relate them to each other. These, then, enabled others to follow in the footsteps of the early explorers, by providing detail information of landmarks, hazards and distances, for both military and commercial ventures. Armed with information covering distance and direction, plus any landmarks that one would expect to encounter on the journey, a traveller could plan a route between, say, two cities. A seafarer would need charts defining coastlines, which also showed the position of hidden hazards such as shoals, rocks and reefs. The position of hidden known nautical hazards could be determined when within sight of land. However, when crossing the open and featureless oceans, a seafarer could be out of sight of land, and without any fixed point of reference for days at a stretch. To supplement the absence of fixed geographical references, use was made of the magnetic compass, the changing positions of the Sun during the day, and the Moon and stars at night.

At some point in time, maps and charts were overlaid with an imaginary grid structure that became the lines of latitude and longitude used today. As mentioned in the *"Introduction"*, it was well known, even in prehistory, that the Earth was round. Remember Aristarchus from the Greek island of Samos, who wrote in the 3rd Century BC ..."*that the Earth is a ball and it spins*"...?

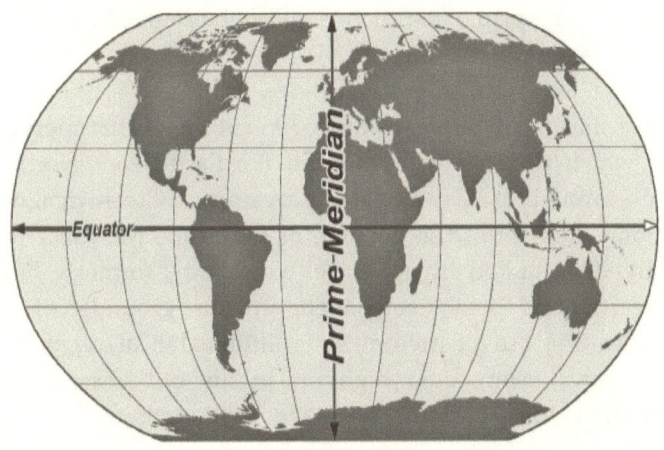

Figure 29   PRIME MERIDIAN

The face of the Earth was divided up into horizontal lines of latitude running parallel to the equator. These lines run from both the equator to the North Pole and from the equator to the South Pole. As there are 360° degrees in a full circle, or 180° in a half circle, the angular changes in latitude range from 0° at the equator to 90° North at the North pole, (also defined as +90°), to 90° South at the South pole, (also defined as -90°). The traveller's latitude, expressed in degrees, defines how far North or South of the equator their position is.

In addition, the face of the globe was divided up into vertical lines of longitude running between the two poles at right angles to the equator. These lines of longitude determine the extent of one's excursion East or West around the equator, again expressed in degrees. However, to have any meaning, it has to be tied to some fixed point of reference. The particular line of longitude running North to South through the point occupied by an individual is termed that person's *meridian,* and the fixed reference line of longitude is called the *prime meridian.* The prime meridian takes on the role of being the 0° longitude.

Today that fixed reference line of longitude is that meridian line which runs vertically around the Earth from North to South through the Old Royal Observatory at Greenwich, near London, England. The position of the Prime Meridian was, until 1999, marked by a plaque, which was, formerly brass, now stainless steel, embedded in the grounds of the Observatory, but at present, it is illuminated at night by a green laser beam sited high up in the old observatory's Flamsteed House building, projected northwards.

The objective was to create a reference meridian line that aligned with the Sun at Greenwich, due South at midday. This was to become known as 12:00 noon Greenwich Mean Time, GMT, or 12:00 Universal Time, UT. Daylight Saving Time, DST, was first advocated in the United Kingdom by George Vernon Hudson in 1895, and was adopted by a number of other countries, but by no means all.

The position of the Greenwich Prime Meridian has been adjusted slightly four times since the 17th Century as measurements that were more accurate were made. A silver plaque embedded in the wall marks the position of the original Greenwich Prime Meridian, made by the first Astronomer Royal, John Flamsteed. The present prime meridian is based on that laid down in 1850-1851 by the seventh Astronomer Royal, Sir George Airy.

However, the Old Royal Observatory at Greenwich was not always the choice of nations for a prime meridian. It will come as no surprise that there have been numerous prime meridians in past times. In antiquity, the Greeks had a line that passed through Paphos, Cyprus, and for a while prime meridians were in place in the USA, Brazil, Canary Islands, Portugal, Spain, France, Belgium, Switzerland, Italy, Norway, Sweden, Denmark, Poland, Romania, Egypt, Russia, India, Japan and Saudi Arabia. This meant that every one of these countries each had to have their own maps and charts, based on their particular chosen Prime Meridian. This obviously had an adverse impact on international commerce and trade. A meeting was held in Washington, USA, in October 1884, to try to standardise on an internationally accepted position. From a delegation of 25 countries, 22 agreed to adopt the British prime meridian. France and Brazil abstained, and San Domingo (now Haiti) voted against the motion. France continued with its own national standard centred on Paris for several decades before falling in line with the rest of the world.

The meridian running down the opposite side of the world is the 180° meridian, which forms in part, the International Date Line. This date line delineates one day from the next, as the Sun appears to travel full circle around the Earth. The International Date Line does not rigidly follow the 180° meridian but deviates in places to accommodate local geographical adjustments for convenience. As the Earth rotates, one side will be in daylight whilst the other side is in darkness, and as such, there are local time zones associated with the Prime Meridian to accommodate this. As the Earth completes one synodic rotation in 24 hours, it means that the Sun is apparently travelling around the Earth at a rate of 15° an hour. Therefore, to facilitate this, the Earth is generally divided into various time zones of 1 hour for every 15^0 of longitude. Once again, there are adjustments to accommodate local geographical and political conditions. The introduction of Time Zones serves to simplify commerce and integrate life styles between peoples of the world separated by significant differences in longitude, which affects their differing local hours of daylight. It enables the various peoples of the world to specify their own particular solar noon for example, when the Sun is at its highest altitude, in their own local time.

Time zones are often named. As an example, the main landmass of the United States is about 2,700 miles wide, ranging from longitude 65° west to 125° west. Hence, with a span of 60° of longitude there are four time zones. Travelling East to West, they are "Eastern Standard Time", "Central Standard Time", "Mountain Standard Time", and "Pacific Standard Time". The Sun will rise in Seattle four hours later than it does in New York, and so by taking account of these time zones it enables peoples living in different locations to conduct their daily business with each other.

Putting this in perspective, when it is 12 noon Greenwich Mean Time or Universal Time in London, England, there will be a lag of 8 hours with the Pacific Standard Time zone, which means that local time in Los Angeles will be 4am in the morning. Such time differences can change by an hour at certain times of the year if either party employs daylight saving times.

This, then, details how the Earth is sectored defining relative locations and times. The same will need to be applied, in some manner, to the Sun.

The Equation of Time

Taken together, both this and the following chapter will show how we can navigate around the surface of a ball of gas and ascribe heliographic coordinates to changing surface features such as sunspots. A major feature of this is the manner in which the rotation of the Sun is defined, in terms of longitude, through the application of Carrington Rotation periods. This is the subject of the following chapter, but first we need to establish a means of defining time at any given location on the Earth and at any point during the day so that we can equate it to the corresponding attitude of the Sun's daily rotation.

There are daily irregularities in the true duration of the Earth's synodic rotation, and in essence, the Equation of Time accounts for these. The Babylonians first noticed this daily variation some 5000 years ago. It is occasioned by two phenomena. The first is due to the slight elliptical nature of the Earth's annual orbital path around the Sun. The second effect originates from the tilt in the Earth's pole of rotation relative to the plane of the ecliptic. These two effects will be considered individually. Altogether, the actual difference in the duration of one synodic day to the next is only a few seconds and since mechanical or electrical clocks and watches are used to gauge time, this goes unnoticed in daily life.

There are four days during the year when the true mean solar day will be exactly 24 hours. These dates normally fall on April 15, June 14, September 2, and December 25, but may vary by a day due to the inclusion of Leap years. The maximum deviations from the norm generally occur on or about February 12 when the "real" length of the day is 24 hours, 14 minutes, 20 seconds. November 3, plus or minus a day, is the shortest day by 16 minutes 23 seconds, being only 23 hours, 43 minutes, 37 seconds long. This means therefore, that on February 12 the Sun arrives due South at the Old Royal Observatory, Greenwich, 14 minutes and 20 seconds late compared to the "clock time" of 12 noon. On the shortest day, November 3, the Sun arrives South at Greenwich early by 16 minutes and 23 seconds at 11:43:37 UT, clock time.

The first of these two effects considered is the eccentricity of the Earth's orbit. As the Earth follows its orbital path towards perihelion, getting closer to the Sun, it will experience an acceleration under the increasing influence of the Sun's force of gravity, and likewise, will slow down as it recedes towards aphelion. The Earth's orbital velocity changes from a maximum of 30.29 kilometres a second, or 67,771 miles per hour, to a minimum of 29.29 kilometres a second or 65,534 miles per hour, a difference of about 2,200 miles per hour. The average velocity of its journey through space is 29.78 kilometres a second, 66,630 miles per hour.

On passing perihelion in early January, whilst still under the influence of the enhanced solar gravitational field, the Earth continues to accelerate for a while although it will have begun its journey away from the Sun. The rate at which it accelerates decreases as the Sun's gravitational field intensity decreases, reaching its maximum velocity in early April. Thereafter it will slow down on its journey towards aphelion, and will continue slowing having passed it, reaching its slowest velocity in early October. The effect this alone has on the length of a day, throughout the year, is shown in figure 30 on the next page.

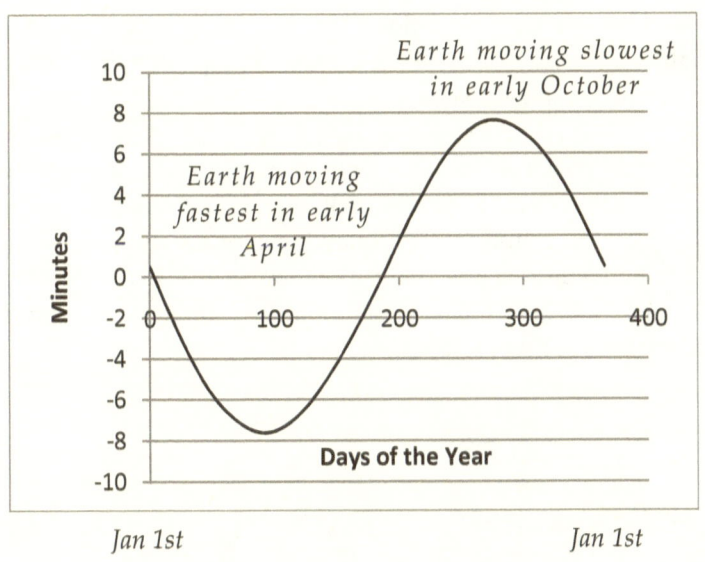

Jan 1st   Jan 1st

**Figure 30   EFFECT OF THE EARTH'S CHANGING ORBITAL VELOCITY ON
THE LENGTH OF THE MEAN SOLAR DAY.**

The graph shows that the duration of the mean solar day
would decrease by 7.66 minutes at its maximum orbital
velocity, and when the Earth is moving at its slowest the
length of the mean solar day would increase by 7.66 minutes.
Once again, the dates quoted will vary by a day or so due to
the inclusion of Leap Years. This contribution to the Equation
of Time correction factors to the 24 hour mean solar day is
solely due to the Earth's changing orbital velocity. The chapter
"How Far Away is the Sun?" explains how astronomers are able
to determine the exact distance between the Sun and the Earth
at any point in time and the precise nature of the Earth's orbit.

By using trigonometry, it is a simple matter to calculate the
Earth's actual velocity at any point in its orbit by taking two
incremental measurements over a short time period. With a

collective set of calculations, or by using more advance mathematics such as calculus, a detailed graph like that shown in figure 30 can be created.

We shall now consider the time changing effect due to the Earth's tilt in its pole of rotation, known as its *"obliquity"* to the plane of the ecliptic. The following ignores the previously discussed effect of the daily variation in the Earth's orbital velocity, and now focuses solely on the influence occasioned by the Earth's tilt of rotation.

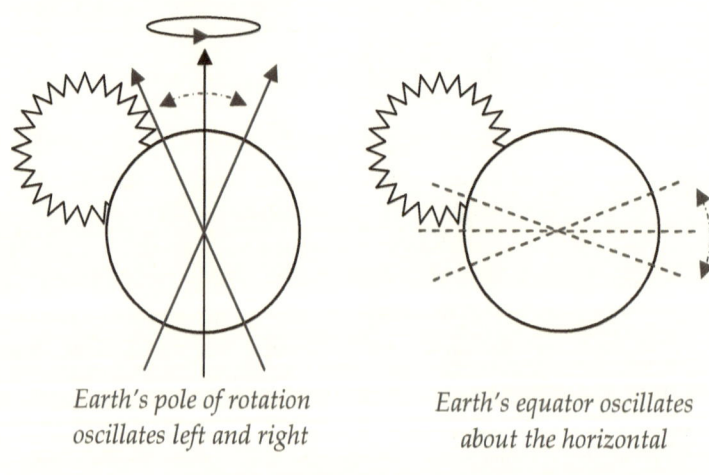

Earth's pole of rotation oscillates left and right

Earth's equator oscillates about the horizontal

Figure 31   THE EARTH'S AXIAL AND POLAR OSCILLATIONS ABOUT THE ECLIPTIC CHANGE THE DURATION OF THE MEAN SOLAR DAY

The 24 hour mean solar day is, in effect, a measure of the rate at which the Sun *apparently* circumnavigates the Earth's equator, or any latitude parallel to the equator. Therefore, the only occasions when the duration of the mean solar day would actually equal 24 hours will be those times when the Earth's pole of rotation is vertical, or orthogonal (at right

74

angles) to the plane of the ecliptic. These occasions occur at the equinoxes when the equator aligns to the plane of the ecliptic and at the solstices when the Earth's pole of rotation is actually vertical despite the fact that it is either inclined directly towards the Sun or directly away from it.

Also on these occasions, the plane of the Earth's equator aligns to the plane of the ecliptic, being horizontal, as in the right hand section of figure 31. In between these events, the Earth's pole of rotation is swinging to either the left or right as the Earth precesses, and the equator cants, or inclines, about the plane of the ecliptic, also shown in the previous diagram.

This has the effect of either lengthening or shortening the mean solar day, by a maximum of 9.87 minutes. The positive going maximum occurs annually at the beginning of May and November, whereas the negative going maximum occurs in early February and August. Figure 32 on the next page illustrates the period of these deviations.

As mentioned, the precise nature of the Earth's orbit and the times of the equinoxes and solstices are determined by observation. Similarly, as shown in the chapter *"How We View the Sun"*, the degree of the Earth's tilt in its pole of rotation, its obliquity to the plane of the ecliptic, is also determined by observation and simple mathematics. From this data, the contribution to the daily variations in the Equation of Time can be calculated. The influence of these two effects has been shown separately, but they combine to produce the overall graph of the Equation of Time daily correction factors. The two components are out of phase with each other and have differing periodicities, but when combined make up the composite response shown in figure 33.

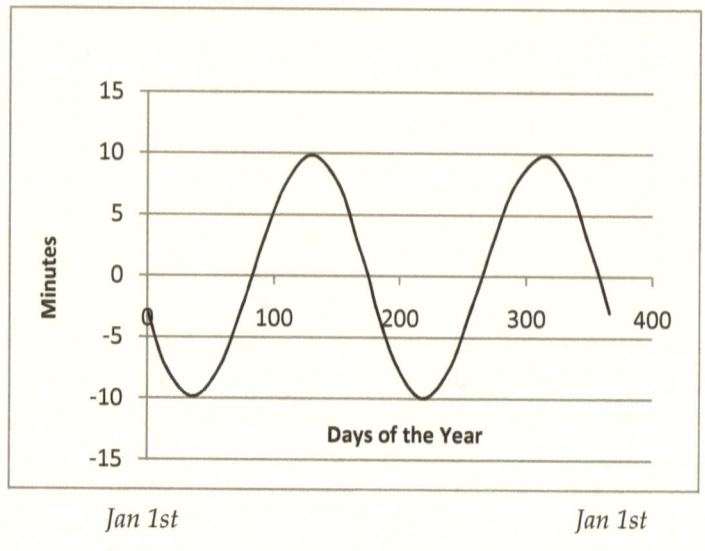

Figure 32   EFFECT OF THE EARTH'S CHANGING POLE OF ROTATION TILT ANGLE ON THE LENGTH OF THE MEAN SOLAR DAY.

The combined amplitudes produce the daily Equation of Time correction factors, shown in figure 33 overleaf. The meaning of the curve is that for the sector of the graph where the time differences are *positive*, the Sun will arrive *early* at the observer's site. Similarly, that part of the curve where the time is negative means that the Sun will arrive late at the observer's site. For example, the maximum negative value of 14 minutes 20 seconds, which generally occurs on February 12, means that this figure has to be added to the time of apparent solar noon at the site, where the apparent solar noon takes account of the location of the observer's meridian.

These daily variations can easily be checked using an accurate clock and a compass to determine the direction of True South for observers in the Northern hemisphere, or True North for those in the Southern hemisphere.

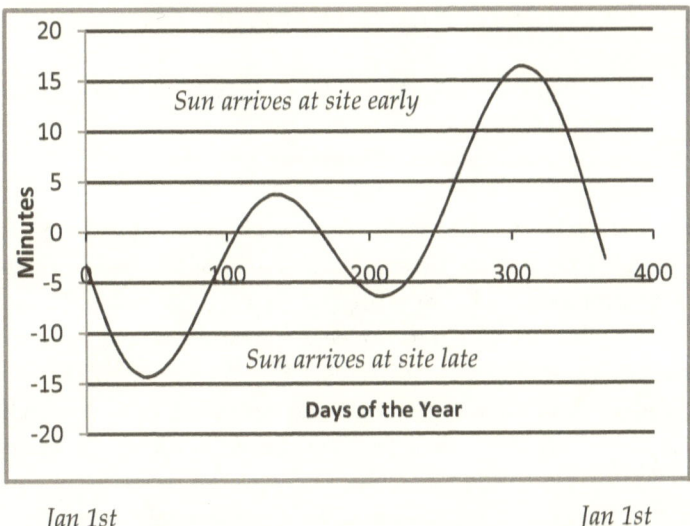

Jan 1st   *Jan 1st*

Figure 33   **COMPOSITE EQUATION OF TIME CORRECTION FACTORS**

The value of the Equation of Time, (EoT), correction factor for any day of the year can be calculated using formulae that describe the curve in figure 33. The main formula is a little complicated and so has not been included here, but for those wishing to perform some calculations and who are familiar with trigonometric functions, an *approximate* formula is offered which is accurate to around ±1 minute:---

$$\text{EoT} = 9.87\sin 2B - 7.53\cos B - 1.5\sin B$$

where $B = \dfrac{360(N-81)}{365}$

and N is the number of the day of the year.

77

Note: "Cos" is an abbreviation of the trigonometrical function "cosine", which is defined as the angle relating to the ratio of the the length of the adjacent side of a right angled triangle divided by the length of the hypotenuse, (the longest side).

To take account of Leap Years, more accurate values of EoT parameters have been used in this text, averaged over a four year period. A complete list is provided in *Appendix 3*.

The correct Local Solar Noon at any observer's longitude or meridian can now be determined. Geographical time zones are accommodated by referring all time to GMT, or UT, but local Daylight Saving has to be considered where relevant since it is not in use worldwide. Consider that solar noon occurs at 12:00:00 UT at the Prime Meridian at Greenwich, England. Take as an example, an observer located at a site in the South West of England whose longitude is 4.4° west. As the Sun apparently completes an East to West 360° orbit of the Earth every 24 hours, it is moving at a rate of 15° an hour, or 1° every 4 minutes. Therefore, it will take an additional (4 x 4.4) minutes to reach the observer's site, which is equal to 17 minutes 36 seconds. This means that the solar noon at the site would then occur at 12:17:36 UT. The Equation of Time correction factor now has to be applied. For this, we shall assume the date is August 6. The correction factor for this date is -5 minutes 53 seconds. The MINUS sign means that the Sun will be LATE and so this figure will have to be ADDED:---

$$12:17:36 + 00:5:53 = 12:23:29 \text{ (UT)}$$

In this example, therefore, true Solar Noon occurs at that site, and on that date, at 12:23:29 UT when the Sun will be due South at its maximum elevation.

The United Kingdom is one of those countries that adopts Daylight Saving, known locally as British Summer Time, (BST), and is in force on August 6. Therefore, another hour has to be added to give local *clock time*. Now, the time displayed by local clocks becomes:---

1:23:29 pm BST

Lines of longitude all pass North – South, crossing the equator, so the Equation of Time correction factors apply equally for locations in both the Northern and Southern hemispheres, as can be demonstrated in the following example.

An observer in Sydney, Australia, is located approximately at longitude 151° east. Relative to Noon at Greenwich, England, the time difference at Sydney, with the Sun moving at a rate of 4 minutes per degree of longitude, will be (151 x 4) = 604 minutes, or 10 hours 4 minutes. So local noon at Sydney will occur early at 12:00:00 UT minus 10 hours 4 minutes, or 01:56:00 UT, since it lies east of the Greenwich Prime Meridian and the Sun moves from the East to the West. As Sydney, Australia is in the Southern hemisphere, the Sun will be situated True North at the local solar noon. Refer to the chapter on *"How We View the Sun"*.

The Equation of Time correction factor for a chosen date of October 20 is +15 minutes 02 seconds. Being a positive value, it means that the Sun arrived *early*, at 12:00:00 *minus* 00:15:02, or 11 hours, 44 minutes, 58 seconds Local Standard Time. Although the solar time difference is 10 hours 4 minutes, Sydney sits in the International 10 hour time zone so local clock time is declared as being exactly 10 hours ahead of GMT, or UT. This means that the observer's wrist watch will read 4 minutes early, 11 hours, <u>40</u> minutes 58 seconds. Some

Australian States do not use Daylight Saving, whereas others do. Daylight Saving is in force at Sydney on that date and so another hour has to be added to reveal that the local clock time of the true solar noon at the site is 12 hours 40 minutes 58 seconds. Equating this time to the time in the UK, the Equation of Time correction factor must also be applied to the UK local time, 01:56:00 UT less 15 minutes 02 seconds, giving 01:40:58 UT in the morning of October 20. Consideration must also be given to UK Daylight Saving, or British Summer Time, which is in operation until the last Sunday in October, so another hour must be added to get the UK clock time.

Putting this all together, solar noon on this date occurs at Sydney, Australia at 12 hours 40 minutes 58 seconds local clock time, when it is 01:40:58 UT in the morning, or 02:40:58 am BST, in the UK.

Astronomers report events worldwide in Universal Time, UT, to provide a level base line on which all data can be compared, and apply the EoT corrections, time zone and daylight saving criteria to their local clock time in order to achieve this.

Carrington Rotations

We have seen from the previous chapter that the true duration of any actual solar day differs from its predecessor and successor due to the behaviour of our own planet and its journey around the Sun. This day to day difference can be as little as 1 second or as much as 30 seconds and generally goes unnoticed in everyday life. It sets a baseline, which accurately defines each daily synodic period and enables us to define which meridian we are presenting to the Sun at any point in time with precision. Likewise, we need to be able to define which meridian the Sun is presenting to the Earth.

It was shown in the chapter *"How Do We Know the Sun Rotates?"*, through the analysis of numerous sunspot readings taken over centuries, that the Sun's equator oscillates North – South through an angle of ±7¼° each year and it has a differential rotation with an average synodic period at latitude 26° of 27.2753 days. This latitude was chosen since it is where sunspot activity is most prevalent through the course of the solar cycle. The 27.2753 day period is an "average", as seen from the Earth. Due to the Equation of Time variations on our own daily synodic periods, it does in fact vary between 27.198 days and 27.342 days. This data has been continually refined over time as more accurate measuring techniques became available. Modern satellite assisted methods, helioseismic surveys, have provided science with very precise parameters.

Unlike on Earth, the Sun has no fixed surface feature to which a Prime Meridian can be tied, but a Prime Meridian is required nevertheless. This was achieved as follows. Back in 1853, during the evening of November 9, at 9.33pm, the noted

English astronomer Richard Carrington arbitrarily assigned a prime meridian to the Sun declaring it simply with the identity "360°". Thereafter as the Sun rotated, this figure decreased to zero, when on completion of the rotation the Sun presented the original incident meridian to the Earth once again. At that point in time the meridian is reassigned 360° and the process starts all over again. From the figures above, we see that the Sun is rotating at a rate a little over 0.5° an hour, between 0.5553° an hour to 0.5554° per hour, to be exact. The incident meridian presented to the Earth at any time is called the Carrington Longitude and is denoted as L_0, (pronounced "L" nought). The diagram below shows the principle.

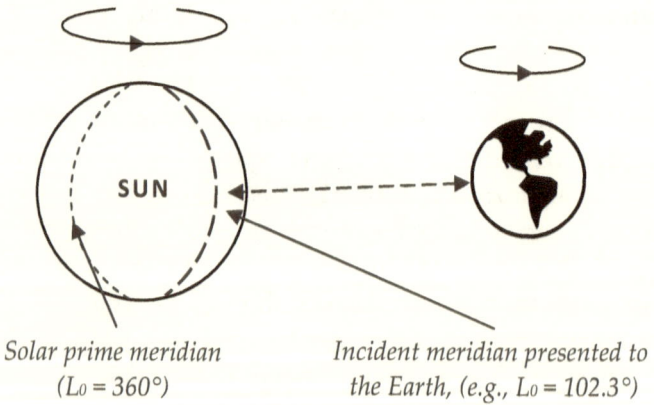

Solar prime meridian
($L_0 = 360°$)

Incident meridian presented to the Earth, (e.g., $L_0 = 102.3°$)

Figure 34   THE CARRINGTON LONGITUDE DECREASES AS THE SUN ROTATES

The incident meridian in this example is $L_0 = 102.3°$. As the Sun rotates anticlockwise about its North Pole, the value of this Carrington Longitude decreases towards zero at a little over half a degree an hour when the solar prime meridian again appears facing the Earth and retakes the value 360°.

Each Carrington rotation is individually identified with a sequential number. As already mentioned, the first Carrington Rotation, CR1, started at 21:33 UT in the evening of 9 November 1853, and it finished at 05:03 UT on the morning of December 7. Its duration was 27.3125 days, or 27 days 7 hours and 30 minutes. CR2 then started at 05:03 UT and terminated at 12:55 UT on 3 January 1854. The duration of this second Carrington Rotation was 27 days 7 hours and 52 minutes, some 22 minutes longer than the first one, the difference being attributed to the variations of the Equation of Time. The Earth was approaching perihelion and its orbital velocity around the Sun was increasing. Each individual Carrington Rotation is tailored to match its particular solar synodic period.

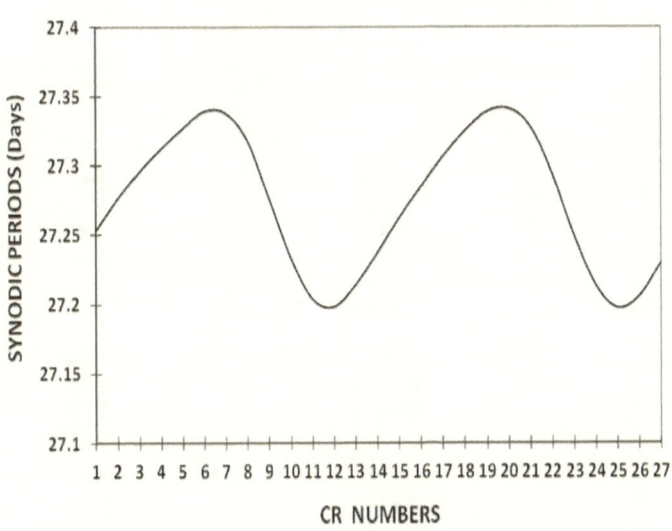

Figure 35   CHART SHOWING TWO YEAR CYCLE OF CARRINGTON ROTATION PERIODS

A further illustration is given in the following table showing a number of actual Carrington Rotations with their associated start/stop times and durations spanning a two year period.

No.	ROTATION NUMBER	START DATE	START TIME	DURATION IN DAYS
1	CR2020	2004-08-18	13:56:14	27.25390
2	CR2021	2004-09-14	20:02:11	27.27964
3	CR2022	2004-10-12	02:44:52	27.29750
4	CR2023	2004-11-08	09:53:16	27.31238
5	CR2024	2004-12-05	17:23:06	27.32708
6	CR2025	2005-01-02	01:14:06	27.33958
7	CR2026	2005-01-29	09:23:06	27.33816
8	CR2027	2005-02-25	17:30:03	27.31856
9	CR2028	2005-03-25	01:08:47	27.27699
10	CR2029	2005-04-21	07:47:39	27.23383
11	CR2030	2005-05-18	13:24:22	27.20426
12	CR2031	2005-06-14	18:18:30	27.19861
13	CR2032	2005-07-11	23:04:30	27.21404
14	CR2033	2005-08-08	04:12:43	27.14193
15	CR2034	2005-09-04	10:01:06	27.27045
16	CR2035	2005-10-01	16:30:33	27.27777
17	CR2036	2005-10-28	23:30:32	27.30650
18	CR2037	2005-11-25	06:51:54	27.32095
19	CR2038	2005-12-22	14:34:04	27.33914
20	CR2039	2006-01-18	22:42:26	27.34159
21	CR2040	2006-02-15	06:54:19	27.32841
22	CR2041	2006-03-14	14:47:14	27.29396
23	CR2042	2006-04-10	21:50:32	27.24969
24	CR2043	2006-05-08	03:50:05	27.21331
25	CR2044	2006-06-04	08:57:15	27.19775
26	CR2045	2006-07-01	13:42:01	27.20625
27	CR2046	2006-07-28	18:39:01	27.23029
28	CR2047	2006-08-25	00:10:38	27.26023
29	CR2048	2006-09-21	06:25:22	27.28447
30	CR2049	2006-10-18	13:15:00	27.30132

Figure 36   TABLE OF CARRINGTON ROTATION PERIODS

Note that the dates listed in the table are expressed in the international scientific format *years, months, days* to counter ambiguities arising from the use of various informal standards by different nations.

The current Carrington Rotation number can be found in any Astronomical Almanac, various handbooks, or a number of websites on the Internet. Some software packages used by astronomers to determine the heliographic coordinates of surface features also provide this data.

The longitude L_0 of the incident meridian for any point in time during a particular Carrington rotation can be found by dividing 360° by the particular rotation period and multiplying by the amount of elapsed time since its start. Then that angle is subtracted from its starting point of 360°.

Take as an example the situation where a period of 8.26 days has elapsed from the start of a 27.142 day rotation period. The angle *"consumed"* during the 8.26 days will be:---

$$360° \times \frac{8.26}{27.142} = 109.56°$$

Hence the meridian L_0 will be 360° - 109.56° = <u>250.44°</u>

So, the position of any surface feature on the Sun can now be defined by quoting the Carrington longitude L_0 on which it is situated, together with the accompanying CR number, and its latitude either side of the solar equator B_0.

Surface Features

Before discussing the various features to be found on the face of our Sun, it is worth reviewing the various types of observations that astronomers make.

Some features resident in the uppermost regions of the photosphere are visible in *white light*. White light observations are made using conventional telescopes that have been fitted with a special filter that excludes light in the infrared and ultraviolet regions of the spectrum harmful to human sight. The remainder of the visible spectrum is available to the observer. This technique, however, is unsuitable for taking observations of certain features in the chromosphere itself and the transphotospheric-chromospheric regions.

Astronomers also make observations in the extreme ultraviolet and x-ray regions. Since the Earth's atmosphere absorbs energy at these wavelengths, satellites operating beyond the Earth's atmosphere make these measurements. Both these regions of the electromagnetic spectrum fall outside of the human visible spectrum, which means that the images taken have to be computer manipulated to introduce false colours so the detail can be brought out. Examinations are made of the upper regions of the chromosphere and that of the corona on the face of the Sun's disc by such methods. Coronal regions around and beyond the Sun's disc can be made by a device called a coronagraph. This is a telescope fitted with an occulter disc which blocks out all the light (visible, infrared and ultraviolet), from the face of Sun that would otherwise obliterate any observation by its glare.

Figure 37   VIEW OF THE SUN'S CORONA USING A CORONOGRAPH

The picture above is from the ESA/NASA SOHO satellite. It is a coronagraph from the LASCO C2 experiment, (Large Angle Spectrometric Coronagraph), taken on 23 December 1996 and shows the Sun's corona with the glare from the Sun's disc blocked out by the occulter disc (in the centre). This picture shows the corona reaching out into space by some 8.4 million kilometres, or 5.25 million miles. To the left can be seen the track of Comet SOHO-6 being pulled into the Sun by its gravity. This image demonstrates the effect that is achievable by employing an occulter disc. Originally, this image was taken in extreme ultraviolet light and has been computer processed to provide a visible record.

Astronomers also observe the Sun with telescopes fitted with filters designed to accept the emissions from particular chemical elements such as hydrogen, calcium, iron and nickel. Under certain conditions, chemical elements emit electromagnetic radiation, some of which is in the visible light region of the spectrum. These emissions occur at wavelengths that are specific to the particular element and the science is explained in the chapter *"What is the Sun Made Of?"*

The Sun is predominately composed of hydrogen and the branch of the science used to look at phenomena resident in and composed of this gas is called *hydrogen-alpha* or H-alpha for short. The name refers to the hydrogen *alpha line* emission that occurs at the red end of the visible spectrum at a wavelength of 6563 Angstroms, or 656.3 nanometres. (A nanometre is one billionth of a metre, or 10^{-9} metre and there are 10 Angstroms to the nanometre). Again, this is explained in the chapter *"What is the Sun Made Of?"*

Calcium, highly ionised by the Sun's magnetic field, emits visible light at the blue-violet end of the spectrum at a wavelength of 3933.7 Angstroms, or 393.37 nanometres, called the calcium-k line. Since calcium is ionised by high intensity local magnetic fields, it makes calcium-k observations particularly useful at providing data about the Sun's magnetic signature in the chromosphere.

Both of the above observational techniques are available to the amateur astronomer as well as the professional.

There are a number of nickel and iron emissions, both within and beyond the visible spectrum, monitored mainly by professional scientists for a variety of reasons ranging from thermal studies of the corona, to the Sun's helioseismic behaviour. Satellites make many of these observations.

Turning now to the various solar surface features, the first and most obvious would be sunspots. The subject of sunspots is extensive, and, as such, has been assigned its own chapter, and so it is not discussed here.

Spicules. The Sun is in a constant state of turmoil with huge cells of hot gases and plasma continually bursting onto the surface of the photosphere. This is the granulation mentioned in the chapter *"The Structure of the Sun"*. When these granulation cells erupt onto the photosphere they set up shock waves in the thin chromosphere above them, and disrupt the local magnetic fields. Jets of plasma, which are about 500 kilometres or 300 miles in diameter, at temperatures around 60,000K, shoot up into the chromosphere from around the cell boundaries where localised concentrations of magnetic flux reside. Speeds reach between 70,000 and 90,000 kilometres per hour, or 45,000 to 55,000 miles per hour. These jets of plasma are called *spicules*. Lasting for about 5 to 10 minutes, they can reach altitudes between 3,000 and 10,000 kilometres, or 2,000 miles to 6,000 miles. Since the chromosphere is reckoned to be between 2,500 kilometres (NASA) to 10,000 kilometres thick, spicules can punch right through the chromosphere ejecting plasma into the highly rarefied corona and on, out into space, to become part of the solar wind.

The surface of the Sun is estimated to have around 100,000 spicules on it at any one time. They are seen on the disc itself as dark spots or tubes, or more easily around the Sun's limb as numerous little spikes. Due to the nature and composition of spicules, and their prevalence in the chromosphere, they are viewed terrestrially using high magnification H-alpha telescopes or exo-atmospherically in extreme ultraviolet light.

Figure 38   SPICULES

The image above was taken on 8 August 2008 by one of the NASA Solar Terrestrial Relations Observatory, (STEREO), satellites and shows spicules on the Sun's disc and limb.

<u>**Solar Flares**</u>. There are other, more violent eruptions from the surface of the Sun. Called solar *flares*, these explosions are generally associated with, and triggered by, the high intensity magnetic flux concentrations associated with sunspots. Flares eject superheated plasma, ionised particles, with temperatures of tens of millions of degrees, which punch through both the chromosphere and corona at speeds that can approach the velocity of light. They generally last only a few minutes and rarely more than 10 minutes. During periods of low solar activity there may only be one a week, or possibly none at all,

but this can rise to several a day during periods of high sunspot activity.

A solar flare also emits electromagnetic radiation spanning the spectrum from radio waves to gamma waves.

Their effect on the Earth is directly linked to the severity of the event and as such, solar flares are rated and classified. Amateur astronomers can observe these events using H-alpha telescopes and grade them in terms of duration, brightness and area. Area can be expressed in square degrees (latitude x longitude) or in "millionths" of the area of the whole solar disc. The BAA has adopted an alphanumeric classification system taking, for example, the format *"1F, 2N or 3B"*, where the number defines the area and the letters quantify their intensity as *"Faint, Normal,* and *Brilliant"*.

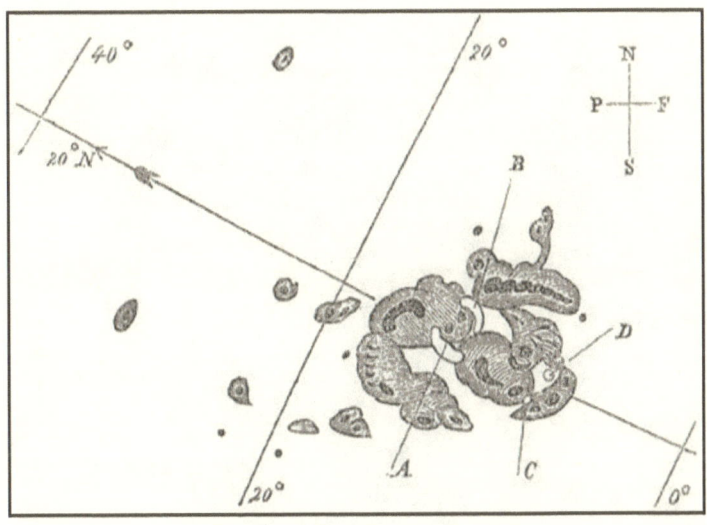

Figure 39   RICHARD CARRINGTON'S 1859 DRAWING OF A SOLAR FLARE

Very occasionally, a solar flare can be so massive that it becomes visible in white light. The first one recorded was during the solar storm (called the solar superstorm) between August 28 and September 2 1859. Two English astronomers Richard Carrington and Richard Hodgson recorded it independently, on September 1. Hydrogen-alpha telescopes were not available in 1859! The previous image, figure 39, is a copy of Richard Carrington's drawing. The flares within the sunspot group are labelled *A, B, C* and *D*.

On the following day, September 2, this solar storm induced electric currents into telegraph lines that burnt out relays and started fires. Two American telegraphists disconnected their batteries and continued communicating for over two hours using the currents induced into the telegraph wires. One was in Boston, Massachusetts, the other in Portland, Maine.

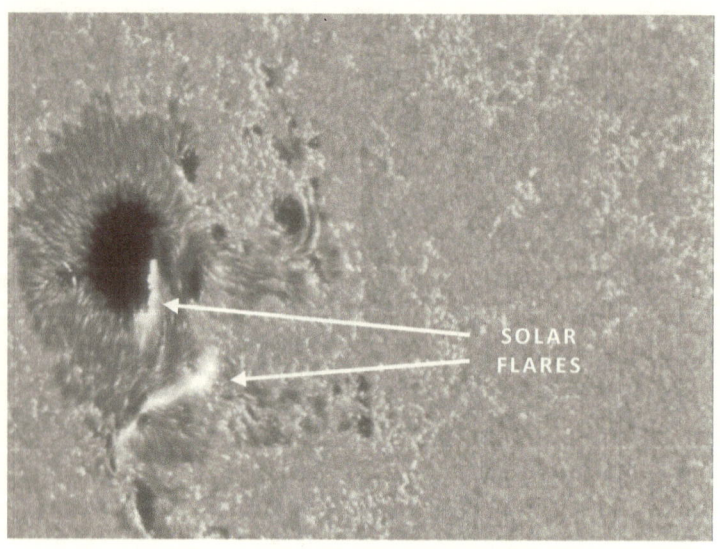

Figure 40   X-RAY IMAGE OF SOLAR FLARES

The previous image is an x-ray picture of solar flares near a sunspot. The Japanese/US/UK spacecraft "Hinode" took it. (*"Hinode"* is Japanese for *"Sunrise"*). The intensity of the x-ray emissions from solar flares is of particular interest, so as to gauge their possible impact on the Earth. These need to be measured exo-atmospherically by satellites, and currently there are a number in orbit above our planet specifically employed for the purpose. Solar flares opposite to, or approaching incidence to the Earth, are of particular interest, since their emissions may be directed towards the Earth. Flares at greater longitudes, or higher latitudes, and those around the limb, do not pose any credible threat because their emissions will be directed out into space and may well bypass the Earth. The photograph below, from the US National Oceanic and Atmospheric Administration (NOAA) Space Weather Prediction Center website, shows emissions from a solar flare directed out into space.

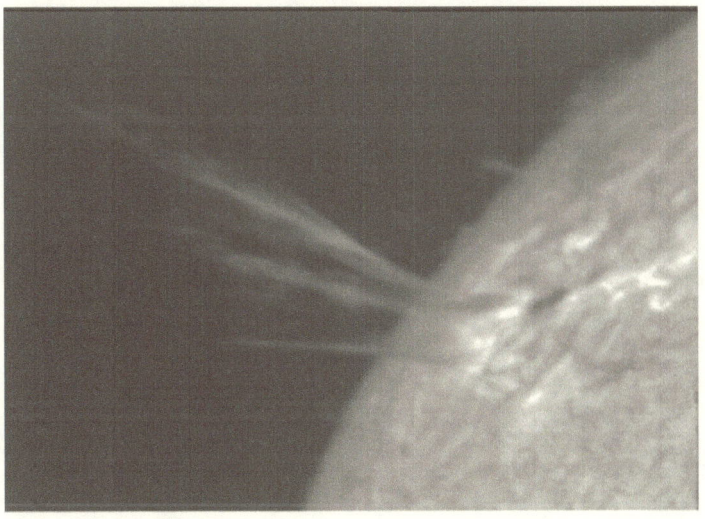

Figure 41   EMISSIONS FROM SOLAR FLARES

Flares are classified into five groups, commensurate with their levels of x-ray intensity, identified as types A, B, C, M and X. The peak radiated flux is measured in Watts per square metre and the x-ray wavelengths are measured in bands between 0.5 and 8 Angstroms.

The most intense is Class X. Flares of this magnitude are major events and can cause widespread, long lasting radio and radar blackouts and radiation storms. Class M is a medium size flare that can cause brief radio and radar blackouts particularly in Polar Regions. A Class C flare is smaller and has only a minor effect on the Earth. Sometimes called subflares, Classes A and B have little or no influence at all. Figure 42 below, of NOAA/SWPC GOES archive data shows x-ray intensity plots spanning a three day period of some major flare activity that occurred in July 2000.

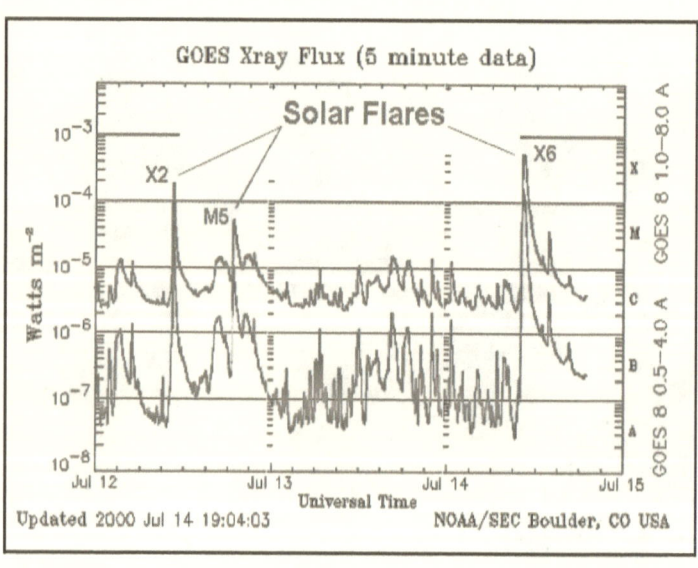

Figure 42 **THREE DAY PLOT OF SOLAR FLARE X-RAY EMISSIONS**

The United States have deployed a number of Geostationary Operational Environmental Satellites (GOES) in geosynchronous orbits above the Earth. Positioned at an altitude of 22,300 miles, (35,800 kilometres), their orbital parameters match the Earth's rotation, so seen from the Earth they always appear to be in the same place. The data used to create the chart in figure 42 was sourced from GOES 8. The chart shows two traces and the right hand scale indicates the lower trace relates to x-rays in the 0.5 to 4.0 Angstroms band and the upper trace 1.0 to 8.0 Angstroms. The left hand scale gives the flux density in Watts per square metre.

The classes are subdivided logarithmically into units of 10, in terms of their radiation flux in Watts per square metre, as follows.

Class A flares, less than 0.1 microwatt/m^2.

Class B flares, between 0.1 and 1 microwatt/m^2.

Class C flares, between 1 and 10 microwatts/m^2.

Class M flares, between 10 and 100 microwatts/m^2.

Class X flares, greater than 100 microwatts/m^2.

Three flares are identified in the chart as M5, X2 and X6. The class M5 flare has a flux density of 5×10^{-5} Watts per metre^2, or 50 microwatts/m^2. The X6 flare has a flux density of 6×10^{-4} Watts per metre^2 or 600 microwatts/m^2. In fact, this flare was a major event and triggered what was named the *"Bastille Day Event"*, after the French National holiday that celebrates the storming of the Bastille in Paris, on 14 July 1789. X-ray energy and associated plasma from the flare arrived on Earth some 15 minutes later, travelling at around half the speed of light, causing a planet wide level S3 *(strong)* radiation storm. This level of intensity affects airline passengers and crew subjecting them to the same levels of radiation they would have received with a chest x-ray. With an S3 storm, satellite systems are disrupted and radio transmissions affected, some giving rise to navigational positioning errors.

Note: Radiation Storms are categorized S1 to S5, *Minor* to *Extreme*, predicting the possible terrestrial impact by the incoming solar electromagnetic energy and charged particles.

Coronal Mass Ejections. There are occasionally even fiercer eruptive events called *Coronal Mass Ejections, CMEs* or occasionally *Coronal Transients.* They are eruptions that can be of cataclysmic proportions, associated with sunspot groups in the same way as solar flares are.

Figure 43 on the next page shows a Coronal Mass Ejection which took place on 4 January 2002, and is a composite picture taken in extreme ultraviolet light by the LASCO C2 experiment aboard the ESA/NASA SOHO satellite. It clearly demonstrates the magnitude of these events. LASCO C2 measurements taken over a period of time have shown that in general these emissions occur at velocities between 20 kilometres per second to 3,200 kilometres per second, (km/s) or 45,000 to 7 million miles per hour, (mph), with an average speed of 489 km/s, or 1 million mph. The average mass of ionised material ejected is estimated to be about 1.6 billion tonnes. However, large explosions can be much higher, equalling the mass of Mount Everest, and occasional giant emissions can eject ionised material greater than the mass of the entire Earth! But how do we know this? Coronal Mass Ejection progress is assessed by real time monitoring, and a measurement of the proton count per cubic metre of the solar wind quantifies the mass.

Figure 43   A CORONAL MASS EJECTION

The reason why these occur is not fully understood but the theory is that in a sunspot group, where solar flares are active, a magnetic reconnection takes place when two oppositely directed fields combine. Within a sunspot group there will be a number of looping magnetic fields, and if two come close enough and they are of opposite polarities (like poles repel, unlike poles attract), they combine. Resident within these looping magnetic fields, are quantities of ionised particles and plasma, moving under the influence of the fields, and when reconnection occurs, they and all their associated energy is released. However, the jury is still out on this one! The chapters on *"Sunspots"* and the *"The Magnetic Sun"* will provide more detail about the magnetic character of sunspot groups, and the solar cycle.

During periods of low solar activity, there may be only one Coronal Mass Ejection event every couple of days, rising to 5 or 6 a day at solar maximum.

Moreton Waves. Many readers will have watched film of aerial bombing and seen the shock waves from the blast spreading out in an ever-increasing circle. A Moreton wave is much the same thing. They are the blast shock waves radiating out from a Coronal Mass Ejection or a solar flare.

First discovered in 1959 by the US astronomer Gail Moreton using hydrogen-alpha telescopes, they have been subsequently studied by the ESA/NASA SOHO satellite and the NASA STEREO satellites.

These waves, which have been nicknamed *solar or coronal tsunami*, reach heights of 100,000 kilometres (60,000 miles) and propagate at speeds between 500 and 1500 km/s (1 million to 3 million mph). Although predominately in the corona, they leave a signature in the thin underlying chromosphere that can be observed using H-alpha techniques. As the wavefront propagates, it emits soft x-rays in much the same fashion as a *Mexican Wave,* made by spectators at a sports stadium. Consisting of hot plasma and magnetism, these waves are known as *magnetohydrodynamic waves.*

Prominences. Prominences always occur on or near the Sun's limb. They are bright regions of dense gaseous material emanating out from the photosphere or may be suspended above it. Attached prominences can break off and drift off into space. They can take many forms but are often seen as loops, arches, spires or pillars, as shown in the next picture. The Sun's disc has been airbrushed out for clarity.

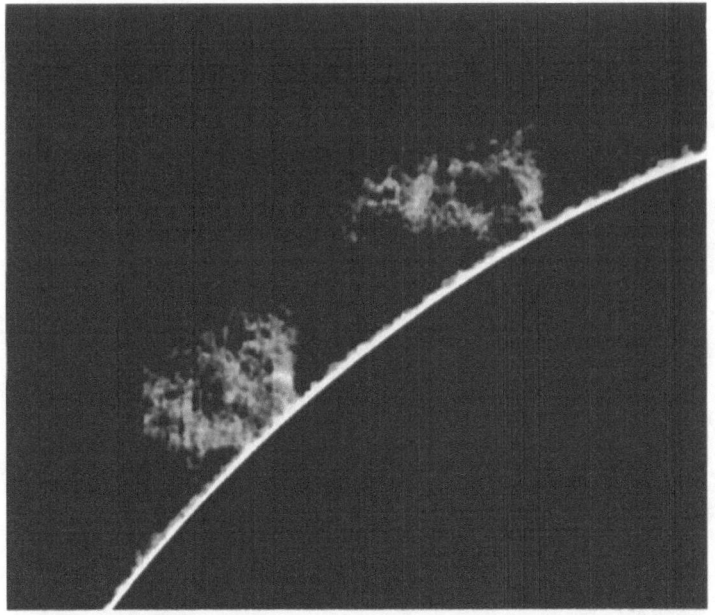

Figure 44   PROMINENCES

Prominences consists of ionized gas and are similar in density and composition to the chromosphere, but are much cooler. They are monitored by satellites using extreme ultraviolet wavelengths and can be seen by earthbound astronomers using H-alpha telescopes. They are visible because they are composed of ionized gas, whereas plasma itself does not emit light. Prominences extend well into the corona and some can be hundreds of thousands of kilometres in length. The one in figure 45, recorded by the NASA Solar Dynamics Observatory (SDO) satellite in December 2010, was 700,000 kilometres (430,000 miles) long.

Small ones can form in as little as an hour, but most take a few hours to a day or so. The majority only last for a few days, but there have been reports of some lasting for weeks and even months, although this is quite a rarity. Figure 45 also has added images of the planet Jupiter and the Earth beneath it to give an idea of scale.

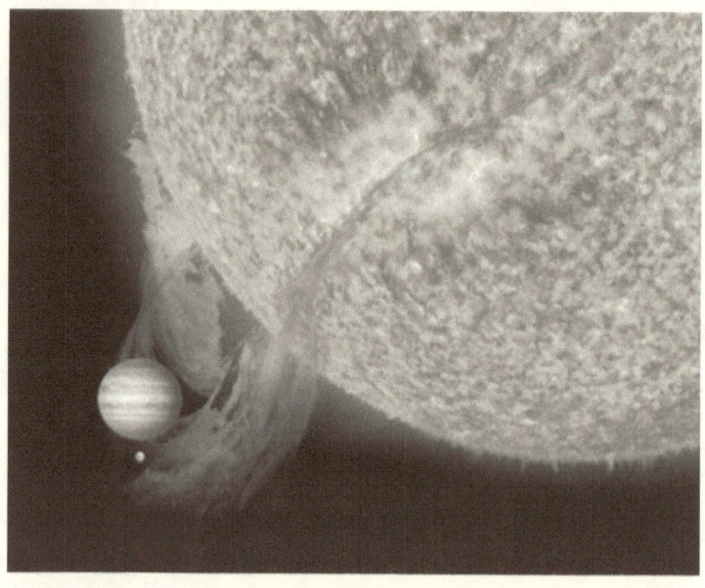

Figure 45 LARGE PROMINENCE WITH JUPITER, AND THE EARTH BENEATH IT, ADDED FOR COMPARISON.

Eventually prominences collapse or sometimes simply explode, possibly giving rise to a Coronal Mass Ejection. Some larger prominences throw out material from the Sun at speeds between 600 to 1,000 km/s (1.34 million to 2.24 million miles per hour). How we know this is explained in the next part of this chapter. Since prominences inhabit the Sun's limb, this material goes harmlessly out into space. Prominences are not yet fully understood and the reason for their birth and deaths are currently unknown.

Filaments. Simply put, a filament is a prominence seen from overhead. The next image is the same as that shown in the previous subsection on prominences.

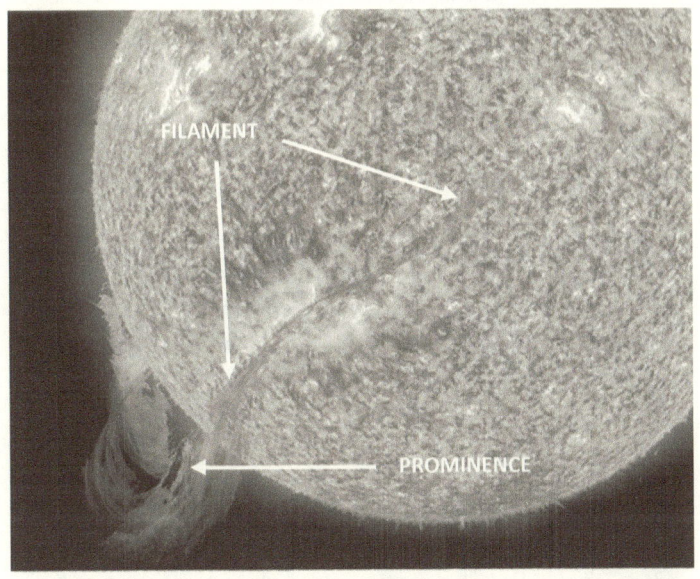

Figure 46   FILAMENTS AND ACCOMPANYING PROMINENCE

The image shows that the filament now appears as a darkened undulating line when viewed against the Sun's disc. As mentioned, it appears dark because it is cooler than the underlying regions. Temperatures are rarely quoted since they can take on a range of values. Since filaments and prominences are much the same thing, these features can be observed terrestrially using H-alpha telescopes or in ultraviolet by satellites.

Filaments, too, may explode, discharging material. This material may be directed towards the Earth and, as such, large filaments pose more of a threat than prominences. With ejection velocities between 1 million and 2 million miles per hour, the material will take 2 to 4 days to reach Earth. With the Sun rotating at around 13° a day at the equator, and less at higher latitudes, filaments from around 20° to 60° East of the Carrington Longitude, L$_0$, will be of interest, as the Earth's orbit could take it into the path of any ejecta.

Faculae. The word *facula* is Latin for *little torch,* faculae is the plural form of the word.

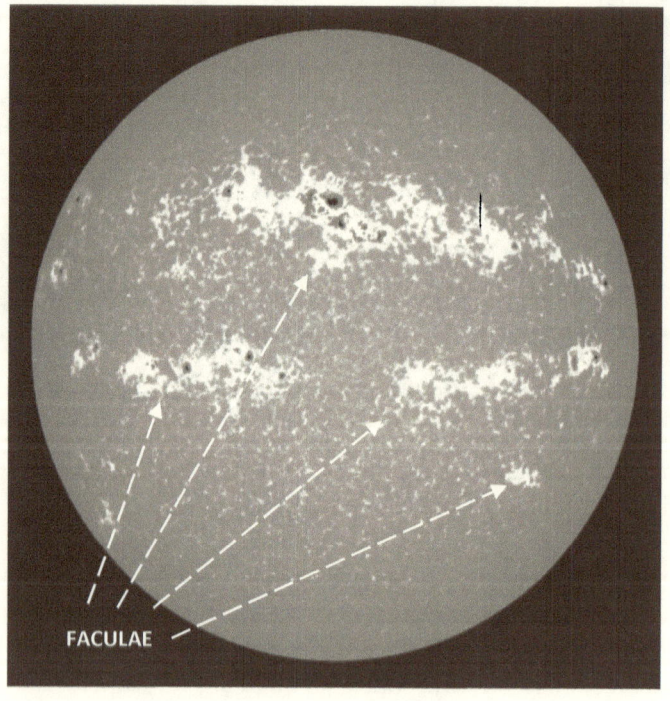

Figure 47   FACULAE

As the name suggests they are embrightened areas resident on the photosphere associated with sunspot activity. Faculae are areas of visible photospheric surface disturbance that may or may not contain sunspots, occasioned by the presence of anomalous magnetic fields. As such, faculae are viewed using white light techniques. This surface turmoil may increase as the intensity of the magnetic activity builds, eventually resulting in the formation of sunspots, or they may linger after sunspots disappear. Faculae may herald the arrival of sunspots or bear witness to their passing, for several days. Faculae can appear and fade without sunspots ever appearing, if the magnetic disturbances fail to reach the required levels for sunspot formation. The image in figure 47, reproduced here with the kind permission of the NASA Goddard Space Flight Center, shows faculae during a period of high solar activity on 28 March 2001. The area appears slightly brighter than its surroundings because the increased magnetic activity raises the local surface temperature of the photosphere. Again, since temperatures can cover a fairly wide range of values, they are not quoted.

Plages. Both faculae and plages appear very similar in black and white photographs, but they are, in fact, quite different phenomena. The word *plage* is from the French for *beach*. Plages, (pronounced *plarj*), are associated with sunspots or active areas and reside above the photosphere, in the chromosphere. They are regions of hot hydrogen and helium gas, held up by ejected material and magnetic fields. This much is known, since the material signatures can be measured using spectroscopy, and the magnetic properties by techniques detailed in the chapter *"The Magnetic Sun"*. They appear as embrightened regions, because the chromosphere surrounding them is cooler, and unlike faculae that are viewed in white light, hydrogen-alpha or calcium-k telescopes are needed to study plages.

Shown here by courtesy of the NASA, Marshall Space Flight Center, the image below is of plages, and was taken using hydrogen-alpha photography on 13 May 1991.

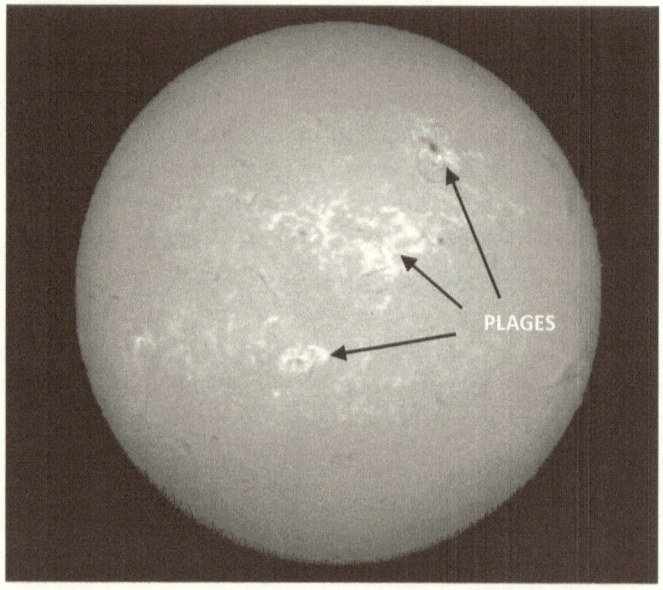

Figure 48   PLAGES

<u>Coronal Holes</u>. As the name suggests, a coronal hole is a hole in the corona, or at the very least, a region of the corona where the local gas density typically falls to about 1% of normal levels. And it should be remembered that the corona itself is a region of highly rarefied gas to start with. The theory is that over an area, the coronal gases have been stripped away by a local anomaly in the Sun's magnetic field.

The left hand diagram below shows the normal distribution of magnetic flux from a bar magnet, representative of a stable solar or planetary magnetic field, and alongside, in principle, an illustration of a magnetic disconnection.

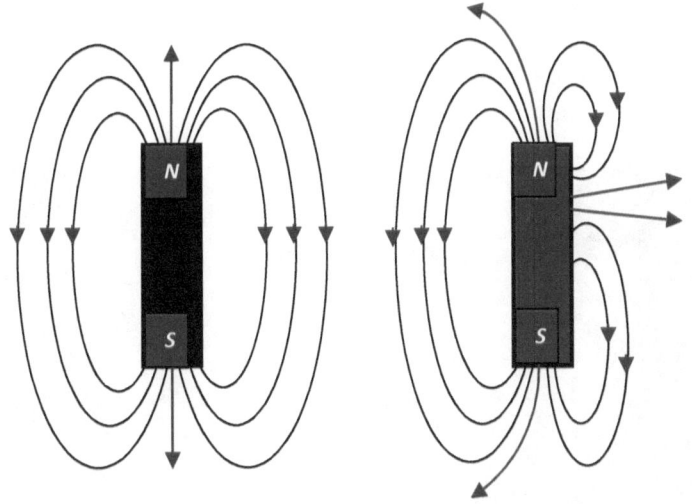

Figure 49   PRINCIPLE OF MAGNETIC DISCONNECTION

Normally the gases surrounding the Sun would be contained by a closed magnetic field whereby emanating lines of flux loop back down to the photosphere. Due to the turbulent nature of the Sun and the various surface influences on its magnetic signature, in reality, the flux distribution around the Sun's surface will be somewhat irregular, and not as idealised as suggested in the illustration.

Occasionally, for reasons not fully understood, a magnetic disconnection occurs and the unipolar concentrated flux in that region extends straight out into space. The local coronal gases are now no longer confined, and the solar wind can escape through the void at around twice its normal speed. Since there is no attenuating influence from the now absent coronal gases, the intensity of the solar wind is higher than it would otherwise have been.

In 1973, the NASA space station Skylab, using a soft x-ray telescope, discovered coronal holes, and today they are observed exo-atmospherically in both x-ray and extreme ultraviolet light.

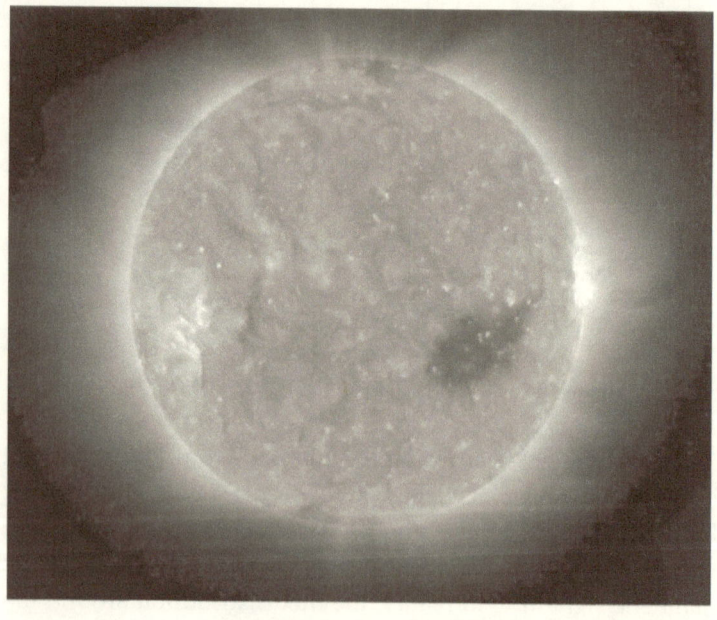

Figure 50   CORONAL HOLE

The NASA STEREO satellite took the image above in extreme ultraviolet light on 25 May 2007. The coronal hole is the void, showing up as the dark patch.

As the Sun rotates, the Earth may traverse the stream of intense solar wind emitted through a coronal hole. In concert with the other traumatic events already discussed, it is at times such as these that the Earth is more likely to produce the majestic displays known as the Northern lights or the Aurora Borealis, together with the Aurora Australis in our Southern hemisphere.

What is the Sun Made Of?

The manner in which the Sun and the Solar System were believed to have been created is dealt with in the chapter *"The Life Cycle of Stars"*. The mother star must house a nuclear furnace in its core, which will be hot enough to produce chemical elements through the process of nuclear fusion. This process can create chemical elements up to atomic number 26, iron, and in so doing, release energy. If the core is sufficiently hot and dense, a few additional elements with slightly higher atomic numbers, such as nickel and cobalt may be produced. On Earth, there are naturally occurring chemical elements spanning the range from hydrogen to uranium, atomic number 92. Two elements are missing. These are technetium, (atomic number 43) and promethium (atomic number 61), due to the fact that they exist only as isotopes with a very short decay time on the geological timescale. It may be assumed that they were present at some time in our prehistory. There are, in all, 109 known elements, the others being created artificially by Mankind. These processes are understood and can be replicated in the laboratory. One such process lead to the invention of the atom bomb!

These same natural elements had to have been present at some point in time in the nascent gas cloud from which both the Earth and our Sun were born. There is only one source in the Universe with sufficient energy where these higher order elements can be created and that is in an exploding supernova! This is the second method by which chemical elements can be created. The outer gaseous regions blown off a supernova, seeded with all the natural elements, can proceed to form a new nebula or merge with another nearby. Thus, it follows that our own Solar System was born from the demise of a supernova. Because the nature of the Sun and Earth are now quite different, the relative quantities of the various chemical elements, after the passage of some billions of years, will also now be different. As the Sun is very hot, some of the heavier chemical elements will break down into less complex elements.

A star's composition changes throughout its life. Primarily it converts hydrogen into helium, and secondly helium goes on to create other elements. Towards the end of its life, dependent on its terminal mass, the core could be composed mostly of iron, and it becomes a ticking time bomb, a supernova!

To appreciate the chemistry further, it is worth revising the basic construction of the atom. For this we shall adopt the classical *Bohr Model* of the atom, named after the Danish physicist Niels Bohr. Surrounding a positively charged nucleus revolves a number of negatively charged electrons keeping the structure electrostatically neutral. These electrons arrange themselves in shells with the first shell containing no more than two electrons. The second shell can contain up to eight electrons, the third up to eighteen and so on. The nucleus is composed of protons, each bound to a neutron. The protons carry the positive charge that balances out that carried by the electrons. In a stable atom the number of protons will equal the number of neutrons and the number of electrons. Remember that an element's atomic number is defined as the number of protons resident in the nucleus.

Should an electron, or even additional electrons, be stripped from an atom, the atom then becomes a positively charged ion.

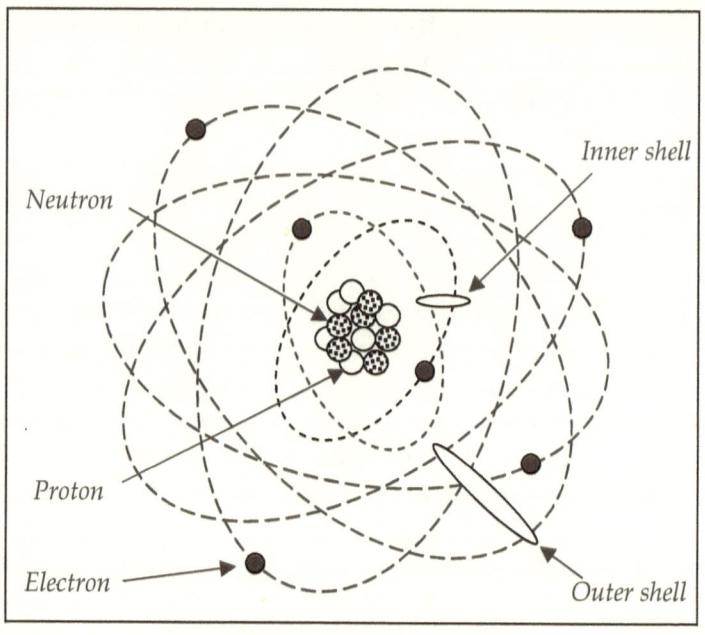

Figure 51 **SCHEMATIC OF A CARBON ATOM**

A stable carbon atom, atomic number 6, has by definition six protons, and therefore six electrons and six neutrons. The inner shell contains two electrons and the next outer shell contains the remaining four. With a nucleus of six protons and six neutrons, it will have an atomic mass, designated C_{12}. Carbon isotopes contain additional neutrons in the nucleus. One such example is carbon fourteen, C_{14}, an isotope with a nucleus of six protons and now eight neutrons, which has a particularly long decay time and is used in radiocarbon dating.

To form a nucleus, a proton must bond with a neutron to form a nucleon. There will be an energy loss, again in the form of an electromagnetic emission. At the same time, the nucleon's mass will decrease. The relationship between mass and energy is well known being Albert Einstein's famous equation, e = mc², and the energy emission due to the mass reduction or *mass deficit*, as it is called, obeys this law as follows:---

$$(\text{change of energy, e}) = (\text{change of mass, m})c^2$$

The graph below shows the relationship between nucleon binding energy, the energy required to separate a bonded neutron proton pair, and the number of nucleons in a particular atom. It is different for every chemical element. As expected it peaks at iron, but since the following fall off is gradual, it suggests that higher elements such as nickel and cobalt could be created in any hotter region within the core.

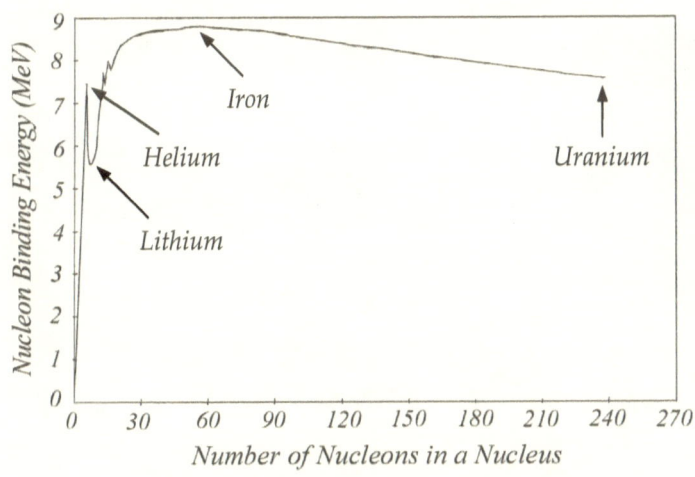

Figure 52   NUCLEON BINDING ENERGY CHART

The binding energy is expressed in Mega-electron Volts (MeV) where one MeV is equal to an energy level of 1.602 x 10⁻¹³ Joules.

Iron has the most stable nucleus due to it having the highest binding energy, and this is why the core of a dying star becomes eventually converted into predominately iron, or possibly iron alone.

In like fashion, an iron atom will have twenty six electrons arranged in four shells. The first shell will contain two, the next eight, the third shell fourteen, with the remaining two in the outer shell. Electrons in each shell will have energies dependent on which shell they occupy. Those in the outer shell will have the highest energy since the electrons orbit at greater distances and at greater velocities. The two electrons in the inner shell will have the lowest energy levels.

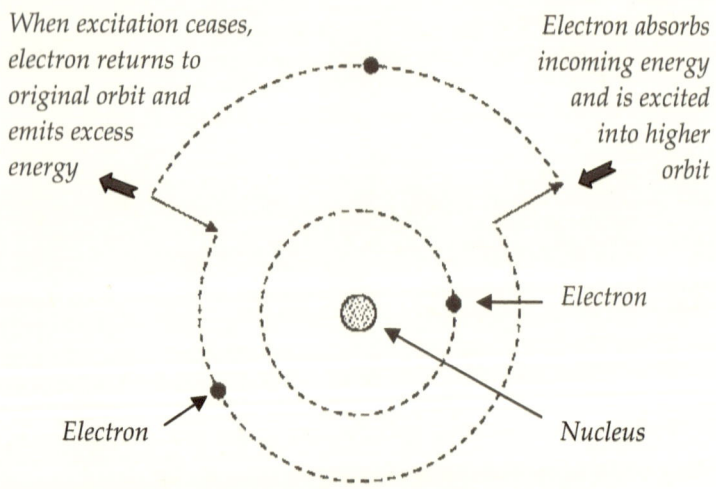

When excitation ceases, electron returns to original orbit and emits excess energy

Electron absorbs incoming energy and is excited into higher orbit

Electron

Electron

Nucleus

Figure 53   DIAGRAM DEPICTING CHANGE OF ORBITING ELECTRON'S RADIUS WHEN EXCITED BY ELECTROMAGNETIC ENERGY

When an atom becomes excited by an external stimulus, such as some form of electromagnetic radiation, one or more electrons will increase their radial orbits and velocities, returning to a relaxed or rest state when the stimulus is removed. When returning to a lower energy level state they will emit their excess energy also as an electromagnetic radiation in the form of light, radio waves, gamma or x-rays.

Below is a simplified schematic diagram of the electromagnetic spectrum, which is not to scale, showing the various divisions from the shortest wavelengths, gamma rays, to the longest wavelengths, radio waves.

Gamma rays	X rays	Ultra violet	Visible light	Infra red	Micro waves	Radio waves

←——— *decreasing*   *WAVELENGTH*   *increasing* ———→

Figure 54   THE ELECTROMAGNETIC SPECTRUM

There is a direct relationship between an energy level and its wavelength in the electromagnetic spectrum....

$$\text{energy, e} = \frac{\text{"\textit{Planck's constant}" x speed of light}}{\text{wavelength, } \lambda}$$

Planck's constant is a fixed number, 6.626×10^{-34} Joules/second, named after its developer, German scientist Max Planck. A Joule is a unit of energy that can be expressed in several ways. Some examples are:---

1 kilogram.metre^2/second^2.
1 Watt/second.
1 Newton.metre.

A Newton is a unit of force equal to 1 kilogram.metre/second^2.

An electron can be knocked out of an atom through a collision with an incident free electron, like a snooker ball hit by the cue ball. In this case, an electron from a higher shell will drop down into the gap, emitting its excess energy in the process.

Now it has been said that the electrons in an atom have different energy levels dependent on which shell they occupy. In addition, they will have different energy levels dependent on the total number of electrons in the atom. The atom of a particular chemical element can jump in and out of many different energy levels, and these various energy levels will differ between different chemical elements. This will happen only at various discrete energy level differences. If an atom has its energy level raised sufficiently, an electron in the outer shell may attain such orbital velocity that it breaks free and flies off the atom. The electron is then said to have entered the *conduction band*. (Control of these processes is the basis of semiconductor physics). By losing an electron, the atom becomes ionised, and with sufficient excitation may go on to lose additional electrons. Due to the Sun's high temperature and magnetic fields, its component elements exist in gaseous form, often highly ionised.

The various electromagnetic emanations radiated by a chemical element undergoing these changes are specific to that element and is a signature which can be used to identity that particular chemical. This signature is called an emission spectrum. Analysis of the composition of the Sun is performed by breaking down visible light from the photosphere into its spectral components and matching them to the known spectra of various chemicals. In addition, satellites undertake spectral analyses in the gamma, x-ray and UV regions since emissions at these wavelengths are attenuated or blocked by our atmosphere. Visible white light shone through a prism is split out into its constituent colours, as shown below.

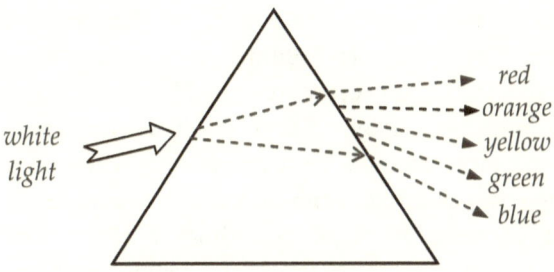

Figure 55   A PRISM SPLITS OUT WHITE LIGHT INTO ALL THE COLOURS OF THE VISIBLE SPECTRUM

The light from the photosphere is separated out in a similar fashion in a spectroscope and the displayed colour lines compared to those of known chemical elements. The relative light intensities between the various elements are used to determine the abundance of the various components. Modern science is able to do this with considerable accuracy. Below is a simplified visible light emission spectrum of the hydrogen atom with its associated wavelengths.

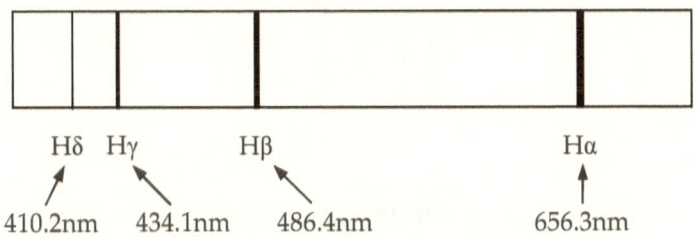

Figure 56   VISIBLE EMISSION SPECTRUM OF A HYDROGEN ATOM

The wavelengths are given in nanometres (nm). Greek suffixes are used to annotate the individual spectral lines, and these spectral lines are known as a *Balmer Series*. The visible spectrum roughly spans that part of the electromagnetic spectrum from 380nm to 760nm. As such, the Hα line will appear red, the Hβ line cyan/blue, the Hγ line bluish violet and the Hδ line deep violet. Readers will be familiar with the bright red glow emitted by a neon lamp when an electrical discharge is passed through it. Neon, for example, has a number of emission lines, and those in the visible range number seven red, six orange, one yellow and one green. Their relative intensities combine to produce the characteristic red coloured glow. Another example, sodium, was used in low pressure vapour street lamps. Sodium vapour lamps emit a distinctive yellow glow centred on 589.5nm, but actually consist of two lines at 589.0nm and 589.6nm, colloquially known as the "sodium doublet".

A spectrum composed of many elements will have an appearance similar to the informal schematic below.

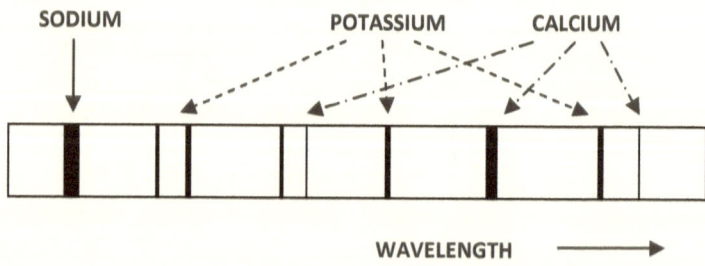

Figure 57   REPRESENTATION OF AN EMISSION SPECTRUM

Emissions outside of the visible spectrum, infrared and ultraviolet, are annotated in like manner, but x-rays are often displayed in terms of their energy levels. A characteristic display may look something like the diagram below.

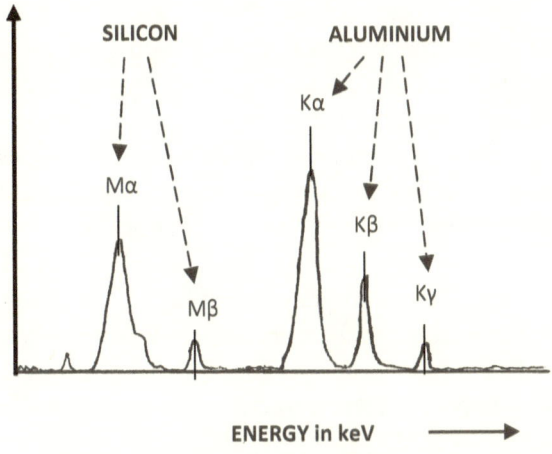

Figure 58   ILLUSTRATION OF AN X-RAY SPECTRUM

These energy levels are expressed in kilo-electron volts (keV) where one keV is equal to an energy level of 1.602×10^{-16} Joules. The "M" and "K" symbols relate to which electron shell in the atom the energy emanated from, and this symbology is called *Siegbahn Notation*.

It was mentioned in the *Introduction* that back in 1814, Joseph von Fraunhofer tried to determine the composition of the Sun. This he achieved, in part, by analysing the Sun's visible light spectrum as outlined above. However, there was one major difference.

We have seen that an atom in an excited state will release energy at a specific wavelength when it relaxes back to a lower state. An atom both absorbs energy and emits it at the same wavelength. Therefore, light passing through a vapour will lose part of its spectrum, absorbed by the particular chemical elements present in the vapour. The resultant depleted spectrum is called an absorption spectrum and is, in effect, the opposite of an emission spectrum. The light from the Sun's photosphere has first to pass through the chromosphere. The gases in the chromosphere will then absorb the energies of their choice. The remaining light travelling to Earth is then the absorption spectrum of the Sun's chromosphere. Fraunhofer analysed this absorption spectrum. He first noticed a number of dark bands in the light from his spectroscope, 574 in all, and set about linking them to the spectra of known elements. These dark lines are still called Fraunhofer Bands today in his memory. From this, he discovered 9 chemical elements. Helium was discovered later through a similar analysis during the solar eclipse of 1868.

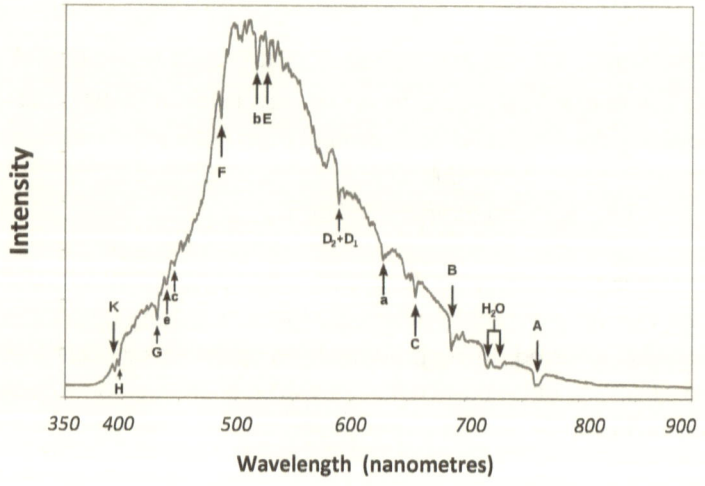

Figure 59   FRAUNHOFER SPECTRUM

The Fraunhofer absorption spectrum shown on the previous page is displayed not as a continuous visible spectrum, but as visible light intensity against the corresponding wavelength. Only a few elements are labelled. The dips in the spectrum occur where chromospheric absorption of the light emitted from the chemical elements in the photosphere takes place. In addition, some absorption occurs in the Earth's atmosphere, and the most notable here are the dips labelled H_2O, water vapour. The table below identifies the Fraunhofer notation used in the diagram above.

Fraunhofer Designation	Chemical Element	Chemical Name	Wavelength (nanometres)
A	O_2	OXYGEN	759.370
B	O_2	OXYGEN	686.719
C	$H\alpha$	HYDROGEN	656.281
a	O_2	OXYGEN	627.661
D_1	Na	SODIUM	589.592
D_2	Na	SODIUM	588.995
e	Hg	MERCURY	546.073
E_2	Fe	IRON	527.039
b_1	Mg	MAGNESIUM	518.362
b_2	Mg	MAGNESIUM	517.270
b_3	Fe	IRON	516.891
b_4	Fe	IRON	516.891
b_4	Mg	MAGNESIUM	516.733
c	Fe	IRON	495.761
F	$H\beta$	HYDROGEN	486.134
e	Fe	IRON	438.355

Fraunhofer Designation	Chemical Element	Chemical Name	Wavelength (nanometres)
G	Hγ	HYDROGEN	434.047
G	Fe	IRON	430.790
G	Ca	CALCIUM	430.774
H	Ca	CALCIUM	396.847
K	Ca	CALCIUM	393.368

Figure 60   TABLE OF FRAUNHOFER ELEMENTS

Using modern, but similar techniques today, the count of chemical elements stands at 69. However, this is the analytical count of elements in the transphotospheric-chromospheric regions and is taken to be the composition of the whole Sun, but what else may be present deep inside?

The curse of Astronomy is the magnitude of numbers!

In order to tabulate quantities from the most abundant down to the smallest trace without clogging up the page with zeros, astronomers have devised a way of presenting this data in an orderly fashion. It is achieved by relating the number of atoms of a particular element present to that of hydrogen. Hydrogen is chosen since it is always the most abundant element in any star, and a table is produced of the number of atoms present of any given element in relation to a trillion atoms of hydrogen.

Astronomers make use of a mathematical process called logarithms. In this case a logarithm (to the base 10) of a number, is that which would have to become a power of 10 to be that number. For example 10^2 is 100, so the logarithm to the base 10 of 100 would be the power of 10, which in this case is 2. Therefore, $\log_{10}100 = 2$. $\log_{10}5 = 0.699$, or $10^{0.699}$ equals 5.

So, a trillion atoms of hydrogen can be written as $log_{10}12$. This means that one can simply put 12 in a table of values instead of 1,000,000,000,000. The convenience of using this system is obvious. It should be noted that in the science of mathematics a logarithm to any base number could be used if desired. Appendix 2 has comprehensive lists of the solar constituents. By tabulating the values as real quantities, the benefit of using this logarithmic approach becomes apparent. A great deal of data can be conveniently held within it whilst avoiding being drowned by all those noughts!

The two major components by far are hydrogen and helium. Many textbooks simply state that the Sun is composed mainly of either 92% hydrogen and about 8% helium, or 74% hydrogen and about 25% helium without explaining that the first set of values are "by volume" and the second set of values are "by mass".

To appreciate the presence of the trace elements, many places of decimals need to be used. For example, it will be noticed that the Sun contains some Promethium, yet on Earth none occurs naturally, due to the fact that it only existed as isotopes with relatively short geological decay times. The amount present in the Sun is only 0.0000000981% by mass, but in reality this actually equates to 1.951×10^{18} metric tonnes, nearly 2 million trillion tonnes. It is produced in the Sun by bombarding Neodymium with neutrons. This produces an isotope, which decays into Promethium in a relatively short time, and in effect "keeps the pot on the boil". It can be seen from the tables in Appendix 2 that there is reckoned to be about 5½ million trillion tonnes of Neodymium still present, left over from that supernova which was the distant ancestor of our Sun.

It will also be noted that there are a number of noble metals present, again in trace quantities. In the figures given in Appendix 2, gold is said to account for 0.000000000766% of the volume, a mere 0.0000001199% of the Sun's mass. But even this amounts to 2.384×10^{18} metric tonnes, nearly 2½ million trillion tonnes, or somewhere in the region of 800 times the mass of Mount Everest! However, some earlier measurements put the quantity of gold at half this amount.

Finally, Thorium is the element with the highest atomic number, 90, accounting for only 0.000000000096% of the Sun's volume. Again this still amounts to 3½ million trillion tonnes of Thorium.

In conclusion, one should not ignore trace elements with values preceded by many zeros, because, due to the size and mass of the Sun, they represent considerable quantities in real terms. The figures given in Appendix 2 were the best available at the time of writing. Other sources quote differing values. One can expect the figures given in Appendix 2 to change and change again in time as measurement techniques and instrumentation becomes ever more sensitive and sophisticated.

To date 69 chemical elements have been detected. Tomorrow there could be more.

Sunspots

Large sunspots can be seen with the naked eye, particularly at sunset or sunrise when the Sun is low in the sky and its glare is curbed by the Earth's atmosphere. Although modern pollution can be an ally, sunspots have been observed with the unaided eye since antiquity, employing methods such as a piece of glass darkened with a film of soot from a candle flame (not recommended), or reflecting an image onto a white surface with a mirror or polished metal. As the name suggests they are spots on the face of the Sun.

It is known that the ancient Chinese had been observing sunspots since the 7th century BC. The Chinese astronomer *Gan De* made the earliest known record in 364 BC. The Benedictine monk *Adelmus* made the earliest record in Western literature on the 17 March 807 AD. Although it was visible for 8 days, he believed it to be the transit of Mercury. If it had been the transit of Mercury, it would have lasted from around 2¾ hours to about 6½ hours, and Mercuric transits occur during the months of May and November, not March. Ooops!

With the advent of the (filtered) telescope, several European scientists first investigated sunspots in 1610 and 1611 but the interpretations put on the observations were frequently incorrect. They were often thought to be planets transiting the Sun's disc. Galileo provided the correct answer in 1612.

Sunspots can be readily observed on Earth using white light techniques, weather permitting! Due to the size of the subject, one can use quite modest telescopes, since very little magnification is required.

Below is a "white light" image of two sunspot groups, taken by the SOHO satellite on 11 March 2011.

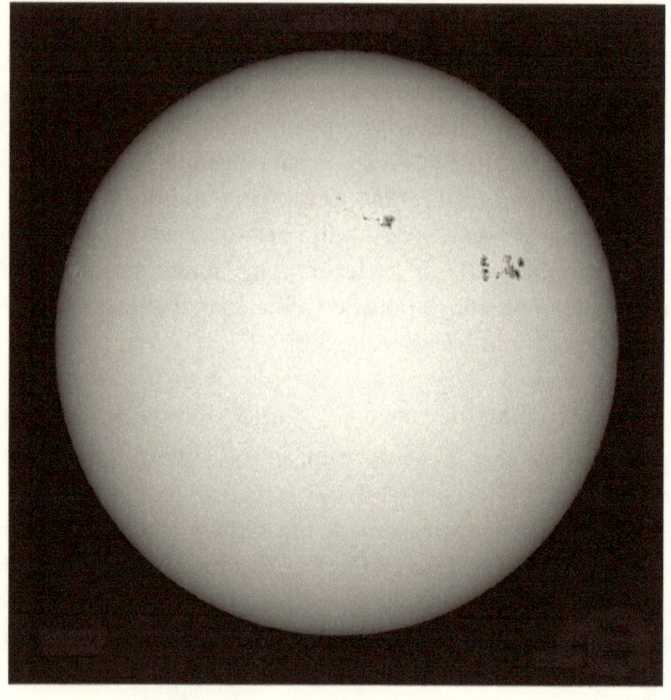

Figure 61   WHITE LIGHT IMAGE OF SUNSPOT GROUPS

Although they encompass all the colours in the visible light spectrum from deepest red to violet, white light images always appear in a greyscale form.

So, what are sunspots? The first thing to note is that they can occur singly or in groups. A single spot can appear and disappear within an hour or may remain for days. A group can appear and disappear within a day or may last for a couple of weeks. A single spot can evolve into a group containing many spots, and a group can develop, increase in size and number, to finally decline and vanish altogether.

The chapter *"How Hot is the Sun?"* demonstrates that a relationship exists between temperature and colour. The *average* temperature of the Sun's surface, the photosphere, has been declared to be 5,776K using this technique. By looking in detail at the true colours in sunspot regions, it can be determined that they are cooler than their surroundings and exhibit temperatures in the range 3,500K to 4,500K. It is the lower temperatures that make them show up as darkened areas against their background in white light images.

Using spectroscopic analysis, they reveal themselves to be regions of high magnetic intensity. This is known by observing the *Zeeman Effect* in the spectrum, which is detailed in the chapter *"The Magnetic Sun"*.

We shall now look at the structure of a sunspot group.

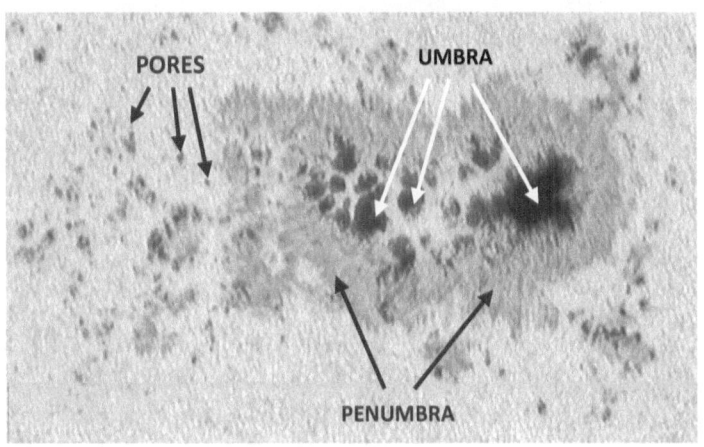

Figure 62   SUNSPOT GROUP

The previous image, taken by the ESA/NASA SOHO satellite, shows a typical sunspot group. Arrows identify the various components. The central darkest area is called the *umbra*, which is Latin for "shade" or "shadow". The umbra, (plural umbrae), is surrounded by a slightly brighter region, called the *penumbra* (plural penumbrae), Latin for "almost shadow". Any number of small spots called *pores* may surround the main structure. We shall look more closely at the umbra and penumbra.

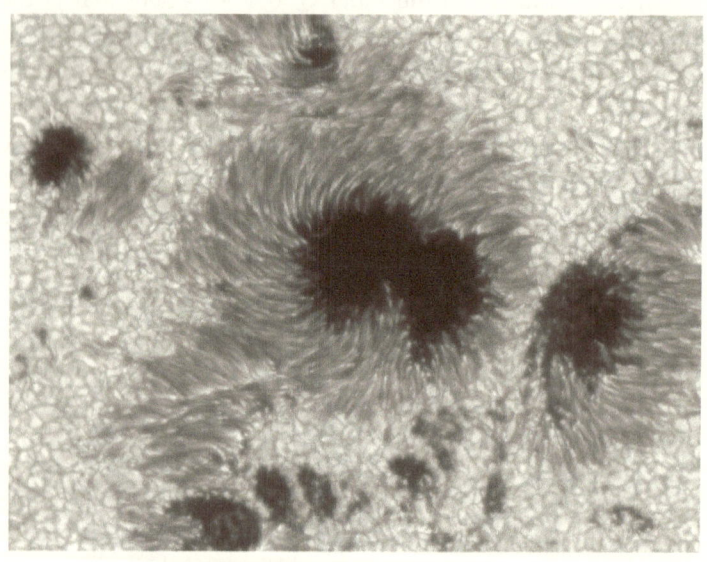

Figure 63   DETAILED VIEW OF MAIN SUNSPOT REGION

The image is of an active area designated AA10030, taken on 15 July 2002 by the Institute for Solar Physics, part of the Royal Swedish Academy of Science. To get an idea of scale, the small spot in the upper left of the picture is bigger than the Earth. Very large sunspots can be up to 80,000 kilometres, or 50,000 miles across, six times the diameter of the Earth!

The image reveals the structure of the penumbra in more detail and shows it to consist of numerous fibre-like features, called *fibrils*. Measurements made by several space missions, notably the Hinode mission, have shown that the fibrils flow material into a positively polarized umbral region from the surroundings, which is carried aloft in the emanating magnetic fields. With negatively polarized umbrae, the direction of flow is reversed. Flow rates vary, ranging from 0.55 km/s (1,200mph) to 18 km/s (40,300mph), where there is a greater propensity for solar flares to occur at the higher transport rates. The magnetic signatures of these features are complex and their influence extends into the chromosphere and beyond. Umbral temperatures are lower than their surroundings since their magnetic fields oppose the transport of convective heat. The next chapter gives more information on the magnetic properties of sunspots.

It is a generally accepted thought that the sunspot umbrae are not flush with the surface of the photosphere but appear to be recessed sunken depressions possibly as much as 1,000 kilometres deep. The Scottish astronomer Alexander Wilson discovered this in 1769. Known as the *Wilson Effect*, it becomes apparent when solar rotation carries sizeable sunspots from central regions towards the edge of the disc, and one is then able to view them "side on". Observations and measurements made around the Sun's limb can be difficult, and so this figure is somewhat imprecise. There are some modern theories that dispute this claiming the feature is an optical aberration occasioned by viewing the sunspot tangentially through a greater distance of photosphere.

We have already seen in the chapter *"How Do We Know the Sun Rotates?"* that by observing the rate at which sunspots traverse the Sun's disc at various latitudes, it has been determined that the Sun exhibits differential rotation. From the numerous observations taken over centuries, additional patterns of behaviour have been uncovered, such as the tilt in the Sun's pole of rotation. In 1843, the German astronomer Samuel Schwabe, discovered that the density of sunspot population follows a cycle. Sunspot activity wanes from periods of high intensity down to possibly no activity whatsoever, then returns to maximum activity. The sunspot cycle duration can vary between 7 and 16 years, taking on average 11 years. At the time of writing, the sunspot cycle is number 24 suggesting that observations have spanned a 250 year period. The Swiss astronomer Johann Rudolph Wolf first numbered the cycles declaring the years between 1755 and 1766 to be cycle 1. Astronomers then reviewed historic records and recompiled the cycles back to Galileo in 1610. However, there is evidence of cyclic records going back in time some 11,400 years.

At sunspot maxima, the amount of UV light reaching the Earth may be 400% higher than at sunspot minima. This variation affected the Earth's ozone layer and the global climate. By studying the growth pattern of tree rings, a science known as *dendrochronology*, the climate for any year is revealed by the annual rate of growth. With a poor climate, cooler years, due to low sunspot activity, growth is slower and the rings are closer together.

Whilst reviewing the records it was discovered that during the 70 year period from 1645 to 1715, there was practically no sunspot activity at all. This occurred during the coldest part of a time known as *"the little ice age"*, when seas and rivers froze over, and fairs were held on the frozen River Thames, London, England. This period of diminished activity has been named the *Maunder Minimum* in honour of the husband and wife team Annie and Edward Walter Maunder who discovered it.

By analysing old records and comparing them with results of low carbon-14 presence, such as periods of low growth in tree rings, other periods of low solar activity have been discovered.

1) The Dalton Minimum, a 40 year period from 1790 to 1830.
2) The Spörer Minimum, a 90 year period from 1460 to 1550.
3) The Wolf Minimum, a 70 year period from 1280 to 1350.
4) The Oort Minimum, a 40 year period from 1010 to 1050.

Due mainly to atmospheric pollution in recent years, the Sun's influence on our weather has diminished markedly.

The average 11 year sunspot cycle still occurs during these periods, but the peaks fail to achieve any magnitude. Although solar activity was unquestionably low, there is one aspect of the Maunder Minimum that should not be overlooked. In the 17th and 18th centuries, the number of astronomers monitoring solar activity was but a few, far less than the army of observers today. Figures quoted for sunspot counts may well have understated the case.

It was discovered at the start of a solar cycle, sunspots tend to appear at mid latitudes between 30° and 40°. As the cycle progresses, they emerge at lower latitudes such that at solar maximum they predominate about the 15° region. This continues and as the end of the cycle approaches, sunspots tend to inhabit equatorial regions around the 7° latitudes. Known as *Spörer's Law*, though first noticed by Richard Carrington in 1861, the German astronomer Gustav Spörer refined and reported on the phenomena.

Another discovery, made by the American astronomer Alfred Harrison Joy, was that sunspot groups tended to have a tilt in their orientation relative to the Sun's East – West line, with their leading edge pointing towards the equator. The angle of tilt approximates to the latitude in which they reside. For example, a sunspot group resident at either latitude 20° North or latitude 20° South, would have a tilt angle of about 20°. Called *Joy's Law*, it is in fact a rule of thumb and is thought to be associated with the magnetic properties of sunspot groups and the underlying mechanism of their creation. This subject will be dealt with in detail in the next chapter. Below is an illustration of the principle, shown here by the kind permission of the NASA Marshall Space Flight Center.

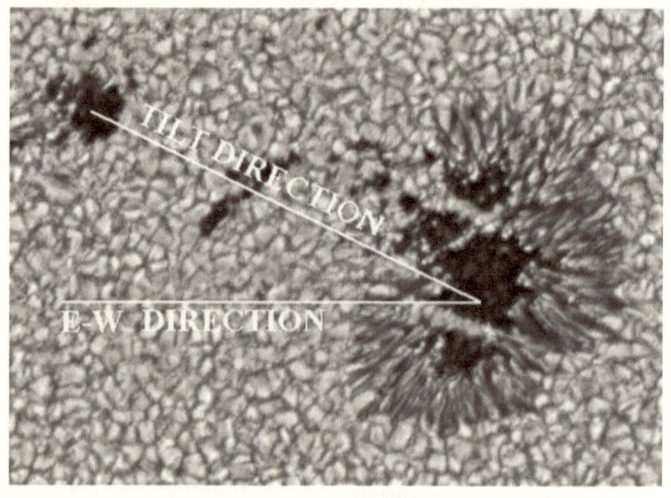

Figure 64   ILLUSTRATION OF THE PRINCIPLE OF JOY'S LAW

Collecting these observations and discoveries, one can create a plot showing the distribution of sunspots and groups over time and at the latitudes they occur. Such a plot is called the *Butterfly Diagram* or the *Maunder Butterfly Diagram*, since Edward Maunder created the first diagram in 1904. Shown below is an example, taken in part from a diagram published by the NASA Marshall Space Flight Center. The displayed data is reminiscent of a series of butterflies, hence the name.

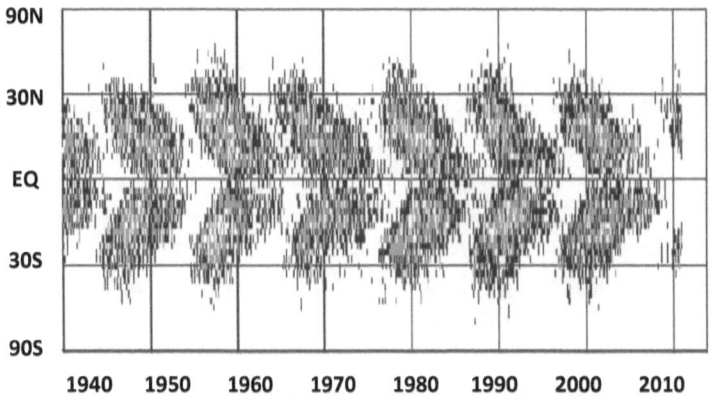

Figure 65   MAUNDER BUTTERFLY DIAGAM

The diagram shows the solar cycles and the manner in which sunspots appear at high latitudes at the start of the cycle, drifting lower as the cycle progresses. Activity at latitudes above 40° does occur, but somewhat infrequently.

Sunspot groups can take many forms and may only consist of a solitary spot with, or without a penumbra. It is useful to categorize groups so as to get a better appreciation of the prevailing solar activity and assess the risk of the occurrence of damaging flares.

Various methods have been tried over the years. One still widely in use today known as *Zurich Numbers* or *Wolf Numbers* is given the symbol R. The Swiss astronomer Johann Rudolph Wolf, from the University of Zurich, first proposed them back in 1848. Quite simply, one adds up all the individual groups

(g) and multiplies the total by 10. To this is added the sum of all the individual sunspots, (f).

$$R = 10g + f$$

This simple formula provides a surprisingly accurate comparative evaluation of the strength of solar activity, yet takes no account of sunspot group structure. Professor Max Waldmeier of the Swiss Federal Observatory, Zurich, proposed the *Waldmeier Classification* or *Zurich Classification* of sunspot groups in 1938. (Do not confuse Zurich numbers with Zurich classification!). Sunspot groups were classified on a scale "A to J", as follows.

Group A   Small single unipolar sunspot.
Group B   Bipolar sunspot group with no penumbrae.
Group C   Bipolar sunspot group. One sunspot must have a penumbra.
Group D   Bipolar sunspot group with penumbrae at both ends. Length of group less than 10° of longitude.
Group E   Bipolar sunspot group with penumbrae at both ends. Length of group between 10° and 15° of longitude.
Group F   Bipolar sunspot group with penumbrae at both ends. Length of group greater than 15° of longitude.
Group G   A great bipolar group with few or no spots between the principle spots. Length of group at least 10° of longitude.
Group H   Unipolar spot with a penumbra, diameter greater than 2.5°.
Group J   Unipolar spot with a penumbra, diameter less than 2.5°.

Though still in use today it has been superseded by the *McIntosh scale* which bases itself on the Waldmeier scale yet adds a two letter code to provide additional identity to the penumbrae and the disposition of the spots within the group. Named after Patrick McIntosh of the US National Oceanic and Atmospheric Administration (NOAA), it was introduced in 1966. As such, the McIntosh scale is also known as the *Modified Zurich Classification*. The McIntosh scale ranges from A to H, omitting G and J, but maintaining the original definitions. With the additional two letter code, Group H simply refers to a unipolar sunspot with a penumbra.

The second, penumbral codes are as follows.

x   No penumbra.

r   Rudimentary, incomplete penumbra partially surrounding the largest spot.

s   Small symmetrical. North-South diameter equal to or less than 2.5°.

a   Small asymmetrical. North-South diameter equal to or less than 2.5°.

h   Large symmetrical. North-South diameter equal to or greater than 2.5°.

k   Large asymmetrical. North-South diameter equal to or greater than 2.5°.

The third, sunspot codes are as follows.

x   This code is assigned to unipolar groups A and H.

o   Opened. The distribution of smaller spots between the principle spots in a bipolar group are few in number or non existent.

i   Intermediate. The distribution of smaller spots between the principle spots in a bipolar group is numerous and none have penumbrae.

c   Compact. The distribution of smaller spots between the principle spots in a bipolar group is numerous and some may have penumbrae. In exceptional cases the whole group may be encompassed in a penumbra.

There are other systems and variations favoured by individual organisations but that above is the most widely used. The purpose of it is to enable astronomers to gauge the likelihood that an active area may produce flares, and to some degree predict the possible magnitude of such flares. Additionally, these classification systems provide a yardstick with which to assess the forthcoming intensity and strength of the solar wind and prepare for an event, if required.

Sunspot groups are numbered for identity. The most widely used system is colloquially known as *Boulder Numbers*. Part of the US National Oceanic and Atmospheric Administration, (NOAA), is the Space Weather Prediction Center, (SWPC), at Boulder, Colorado, USA. The SWPC collects observation data from various US sources such as NASA, US Air Force and others, and publishes a daily *Solar Region Summary* listing the sunspot groups with their assigned numbers, spot count, their heliographic coordinates, McIntosh classifications, sizes and magnetic classification. The web address is *"www/swpc.noaa.gov/ftpdir/latest/SRS.txt"*. The website also provides a full explanation of all the symbols and data used.

It has been stated that sunspots are regions of high magnetic intensity, and that the SWPC publishes the magnetic classification of each group. It has adopted the magnetic classification table created by the Mount Wilson Observatory in Southern California, USA, and is detailed here.

Alpha: A unipolar sunspot group.

Beta: A sunspot group having both positive and negative magnetic polarities (bipolar), with a simple and distinct division between the polarities.

Gamma: A complex active region in which the positive and negative polarities are so irregularly distributed as to prevent classification as a bipolar group.

Beta-gamma: A sunspot group that is bipolar but which is sufficiently complex that no single, continuous line can be drawn between spots of opposite polarities.

Delta: A qualifier to magnetic classes (see below) indicating that umbrae separated by less than 2 degrees within one penumbra have opposite polarities.

Beta-delta: A sunspot group of general beta magnetic classification but containing one (or more) delta spot (s).

Beta-gamma-delta: A sunspot group of beta-gamma magnetic classification but containing one (or more) delta spot (s).

Gamma-delta: A sunspot group of gamma magnetic classification but containing one (or more) delta spot (s).

The Solar Influences Data Analysis Center (SIDC) in Brussels uses the first five categories from Alpha to Delta.

The SIDC is the world centre processing sunspot data, and issues the daily International Sunspot Number, (ISN); a Wolf number derived from observations made worldwide.

The creation of sunspots is dealt with in the next chapter.

The Magnetic Sun

Magnetism, magnetic fields and lines of magnetic flux are invisible, so how do we detect and map them? School children perform an experiment which reveal the lines of flux around a bar magnet by sprinkling iron filings on a sheet of paper placed over the magnet. The filings align themselves to the lines of flux. So, although they are invisible they can be detected by the way in which they affect material objects and matter.

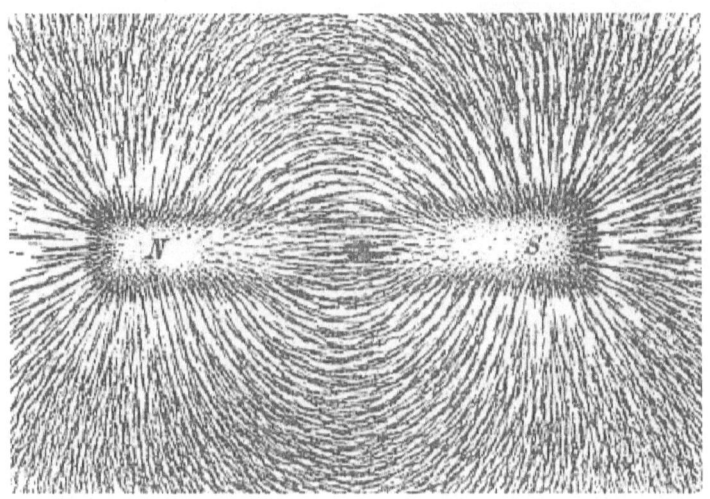

Figure 66   IRON FILINGS BETRAY LINES OF MAGNETIC FLUX FROM A BAR MAGNET

By detecting the ionised material and plasma caught up in them, lines of magnetic flux associated with sunspots can be mapped. Similarly, the Sun's own intrinsic magnetic field can partly be seen by the way it affects the corona.

137

Figure 67   FOOTPRINT OF MAGNETIC FIELD OVER A SUNSPOT

This can be demonstrated by the image above which is of an active region (assigned the Boulder Number AA9373) taken on 21 March 2001 by the NASA TRACE satellite.   TRACE, an abbreviation for Transition Region And Coronal Explorer, was a former mission of the Stanford-Lockheed Institute for Space Research, part of the NASA Small Explorer programme.

The image caption declares the subject to be a "footprint" since magnetic fields are invisible, and what is actually being displayed are the spectral emissions in the ultraviolet (UV) region, around 190 to 200 nanometres, from highly ionised iron atoms present within these fields. By such methods, the magnetic signature can be unmasked.

There is another way of detecting magnetic fields, known as the Zeeman Effect, named after the Dutch physicist Pieter Zeeman who discovered it in 1902. The presence of a magnetic field will cause the spectral signature of a chemical element to alter slightly. In the chapter *"What is the Sun Made Of?"* an explanation was furnished of how an atom can emit electromagnetic energy at various characteristic wavelengths, and so provide a discrete signature of itself, when in transit from a higher energy state to a more relaxed one. Under the influence of a magnetic field, this spectral signature alters slightly. This change occurs in particular to those lines associated with energy level transitions from the electrons in the atom's outer shell.

A simplified explanation of this is that an electron orbiting a nucleus has an energy level associated with its orbital velocity. When exposed to the influence of a magnetic field, the field will either assist the electron's journey or retard it dependent on the polarity and direction of the field. When that electron seeks to transit from a high energy level to a lower one, the penalty or bonus afforded by the magnetic interaction has to be added to the bill. Therefore, the emitted spectral wavelength will have shifted to accommodate the influence of the field. This means three things. First, the shift will be to a longer or shorter wavelength, dependent on the polarity of the magnetic field. Second, that the magnitude of the shift will be proportionate to the intensity of the field. Third, that if a number of electrons are affected, there will be shifts to a corresponding number of spectral lines.

The effect revealed that sunspots were areas of high magnetic field strength. This occurred when attempts were made to perform spectroscopic analysis on the umbral and penumbral regions to see if there was any chemical difference between them. Zeeman splitting has since been used to chart the differing field strengths associated with various sizes of sunspot umbrae.

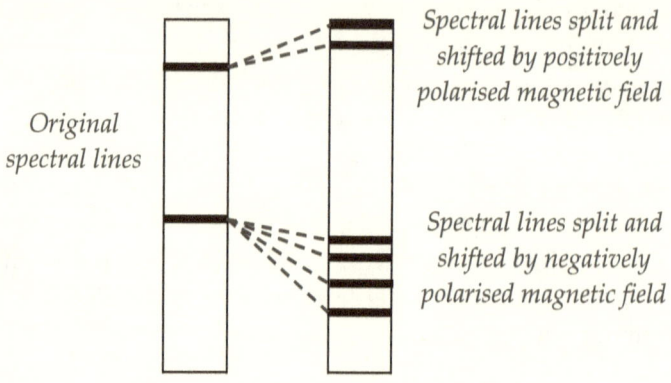

Original spectral lines

Spectral lines split and shifted by positively polarised magnetic field

Spectral lines split and shifted by negatively polarised magnetic field

Figure 68   ZEEMAN SPLITTING OF SPECTRAL LINES

The diagram above illustrates the principle and shows that there may only be one extra spectral line, or there could be many.

NASA maps the photosphere's magnetic anomalies to reveal active regions that may or may not create sunspots in order to get a clearer picture of solar activity. Formerly, this was undertaken by the ESA/NASA SOHO satellite Michelson Doppler Imager equipment and latterly by the SDO Helioseismic and Magnetic Imager (HMI). The SDO/HMI scans at a wavelength of 617.3 nanometres, just off the hydrogen-alpha line 656.3 nanometres. Both missions publish *"magnetograms"* and an example is shown overleaf.

The white areas are positive regions where the magnetic flux emanates out of the surface, and the black areas are regions of negative polarity, where the flux is returning to the surface. Several active areas are apparent, but the magnetogram has the ability to expose active regions that have not attained high enough strengths to create sunspots.

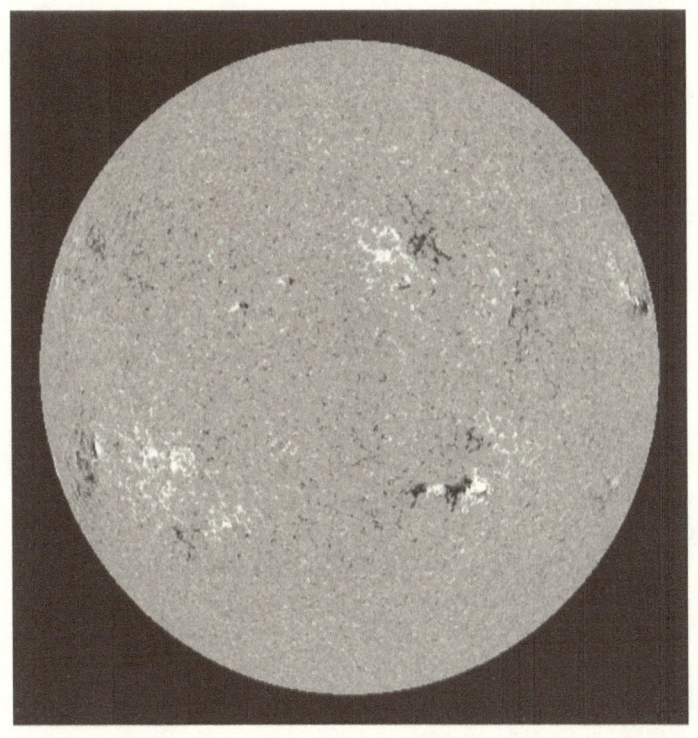

Figure 69   SDO/HMI SATELLITE MAGNETOGRAM

The notion that there could be a relationship between sunspot field strengths, their changing complexity during the solar cycle and the Sun's differential rotation, helped develop the current theory surrounding sunspot creation.   Known as *Babcock's Magnetic Dynamo Model*, the principle is named after the US astronomer Horace Babcock who first proposed it in 1961.

It works like this! The Sun has an intrinsic magnetic field in much the same way as the Earth, and may be considered as a bar magnet.

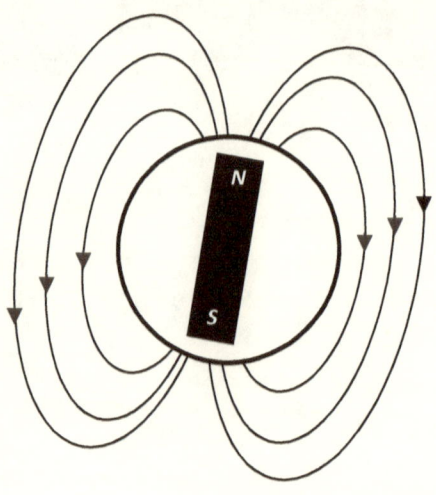

Figure 70   THE SUN'S BASIC MAGNETIC FIELD

The Sun's photosphere is partly composed of plasma, and magnetic lines of flux will move along any plasma flow and so distort the idealistic diagram shown in figure 70. Having been studied under laboratory conditions on Earth, the behavioural links between plasma and magnetic fields are known. It is for this reason that during times of low solar activity, when there is little disturbance of the photosphere and surface plasma flow is at a minimum, the Sun's corona appears "spiky". This is particularly noticeable in the polar regions, where surface emissions are able to follow an undistorted magnetic field.

The Sun's differential rotation moves plasma around the equatorial regions at a greater rate than at higher latitudes. Therefore, the Sun's magnetic fields become distorted as the flux attempts to follow the plasma. Undistorted fields flowing between the poles, the *"initial conditions"* in figure 71, are called *poloidal fields*. As the Sun continues to rotate, plasma flows at different rates, at varying latitudes, and winds up the Sun's magnetic field. These types of lateral fields shown in figure 72, are called *toroidal fields*.

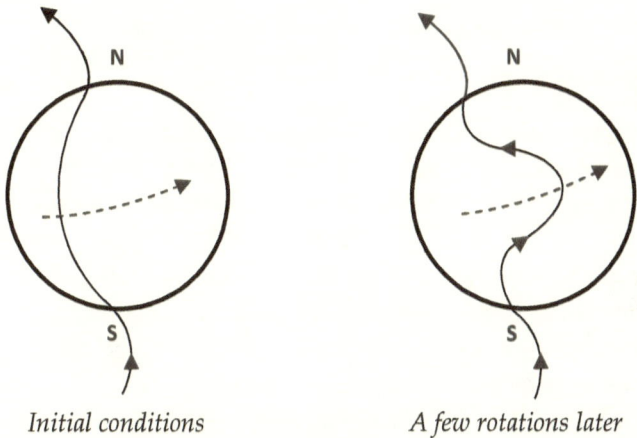

Initial conditions *A few rotations later*

Figure 71 DIFFERENTIAL ROTATION PREFERENTIALLY TRANSPORTS PLASMA AT THE EQUATOR DISTORTING THE MAGNETIC FIELD

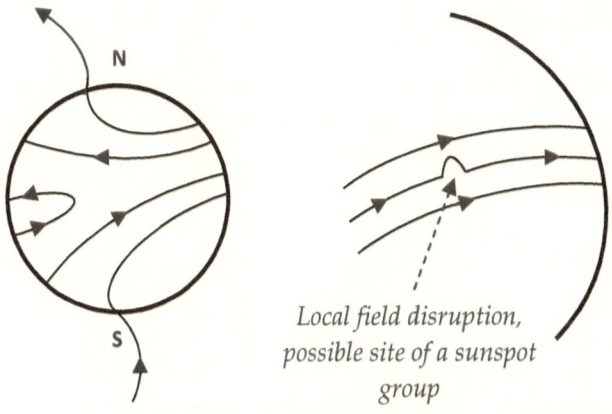

Local field disruption, possible site of a sunspot group

Figure 72 POSITION AFTER MANY REVOLUTIONS. SUN'S MAGNETIC FIELD WINDS UP AND LOCAL DISRUPTIONS OCCUR

The process progresses to a point where it starts to break down. The formation of localised areas of field concentration, which occur due to convection in the photosphere, results in the field becoming locally jumbled and knotted. Where these knots occur, the field breaks through the photosphere and sunspots appear when the field strength is high enough.

In its simplified form, the disruption can be likened to the field from a horseshoe magnet, characterised by the looping lines of flux associated with the sunspot group shown in figure 67.

The convention is that lines of magnetic flux flow *out* from a *north pole* and are given a *positive* polarity. This means that all the sunspot groups in one hemisphere will all be polarised in the same direction, i.e. all pointing East to West, or vice versa All the groups in the other hemisphere will also point the same way, but will now be orientated in the opposite direction. The magnetogram in figure 69 shows evidence of this.

This is known as *Hale's Polarity Law* after the US astronomer George Ellery Hale.

The bar magnet analogy, figure 70, shows that field intensity concentrates at the poles and is a minimum around the equator. Therefore, in the Sun, one would expect field disruption to occur initially at higher latitudes. This is exactly what does happen and can be seen in the Maunder "butterfly diagram", figure 65. As the cycle unfolds sunspot formation occurs at decreasing latitudes, converging on the equator. Since the polarities in each hemisphere are in opposition, when they meet at the equator they cancel each other out. As such, sunspots appearing actually on the equator itself are very rare, and this is why rotational rates given by various authorities are often quoted at higher latitudes. NASA defines the Sun's rate of rotation at latitude 16°, and gives an equation enabling the rate at other latitudes to be calculated, given in the chapter *"How Do We Know the Sun Rotates"*. Equatorial rotational rates have been determined by interpolation.

By this process, the toroidal (east – west) fields become replaced by a regenerated poloidal (polar) field at the end of the 11 year cycle. In addition, when the next cycle begins, the polarity in both hemispheres reverse, returning to their original state at the end of the second cycle. As such, some astronomers consider the solar cycle to be a 22 year event.

The following diagram shows how this works in a simple sunspot group. Such a group with penumbrae on both the leading and trailing edge would be classified as a group D, E or F dependent on the length of the group. The dominant feature here is that returning lines of flux have sufficient intensity to create further sunspots or pores, and possibly sunspots with sufficient strength to develop their own penumbra.

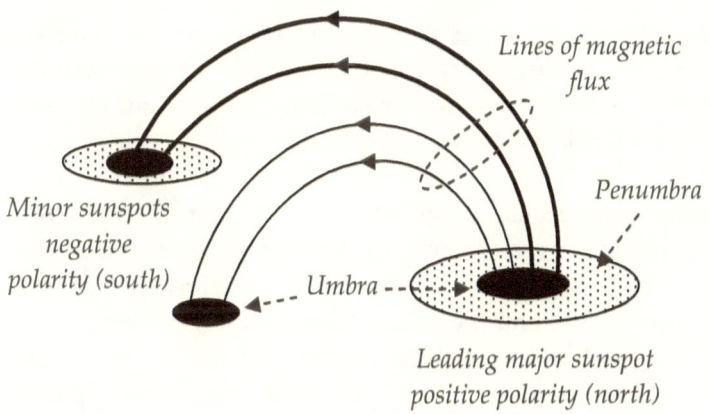

Figure 73   MAGNETIC SCHEMATIC OF A SIMPLE SUNSPOT GROUP

Wherever a dipole relationship exists between the leading sunspot in a group, which is often the dominant sunspot, and the other members, the structure is considered to be a single sunspot group. Only if it can be demonstrated that two (or more) distinct bipolar relationships exist in a complex structure composed of a large cluster of sunspots, that the structure can be considered to comprise more than one group.

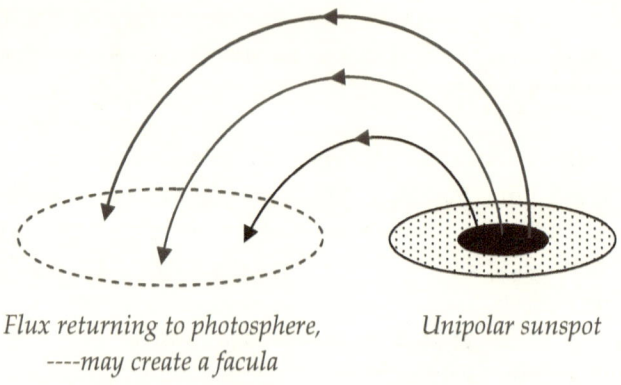

Flux returning to photosphere,   *Unipolar sunspot*
----may create a facula

Figure 74   SCHEMATIC OF A UNIPOLAR SUNSPOT

Unipolar groups, types A and H distribute lines of returning flux over an area of photosphere such that the returning flux density is insufficient to create another sunspot, but may give rise to a facula, as shown in figure 74.

Some professional organisations state that if there is a separation between groups greater than 10° of longitude then they may be counted as two groups. However, sunspot groups evolve, and there have been situations where a group starting out as a class E have expanded into a class F, and beyond, to become falsely counted as two distinct groups where only one bipolar signature was present encompassing the entire structure. It is for these reasons that the US NOAA

Space Weather Prediction Center assign Boulder numbers to sunspot groups in accordance with the data displayed in the SDO (and formerly SOHO) magnetograms.

A lot has been said about sunspots being regions of high intensity magnetic fields. To get an appreciation of field strengths, we must use a unit of measure to describe the magnetic intensities. Anyone studying this subject further will find a number of different units used in literature. They are listed here to show their relationship. The German physicist Carl Friedrich Gauss gives his name to the *Gauss*. One Gauss is equal to 10^{-4} Tesla. Named after the Serbian physicist Nikola Tesla, one *Tesla, T,* is equal to 1 Weber/metre^2, and so on. There are others. It is not intended to define them all but simply use the more popular terms in a comparative manner.

To put things in perspective, a small bar magnet would have a field strength of around 100 gauss, or 10^{-2} T, and a refrigerator magnet about half that. The Earth's magnetosphere has field strengths between 0.31 and 0.58 gauss, (31 to 58 microteslas), whereas the general ambient surface of the Sun has fields between 1 and 2 gauss. Measurements made of large sunspots show they have peak strengths reaching anything up to 4,000 gauss, (0.4 T). In general, field strengths range between 1,000 and 3,000 gauss, but can go as low as 10 gauss. Sunspots come in all shapes and sizes and so cover a wide range of field strengths. The field from a magnetic source decreases as the temperature is raised yet these sunspots are over 4,000K. It should also be remembered that these fields can cover areas greater than that of the Earth! Put together, this represents a vast magnetic environment.

How Far Away is the Sun?

It is about 93 million miles or 150 million kilometres. Well, that's the simple answer; the real answer is a little more involved.

If someone needs to know the height of a building or a tree for example, the traditional method is to use a system called "triangulation". Starting from the base of the tree or building, a distance along the ground is measured out. At that point, as shown below, the angle between the ground and the top of the object is then measured.

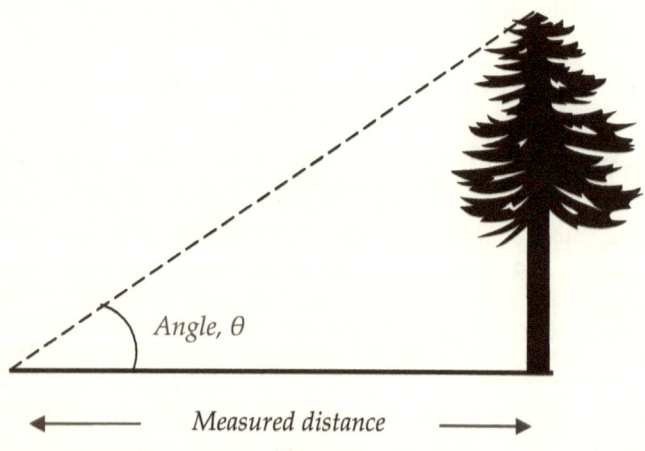

Angle, θ

Measured distance

Figure 75   MEASUREMENT BY TRIANGULATION

There is a trigonometrical relationship between the various sides of the triangle and the angle θ. In this instance, the trigonometric ratio "tangent" of the angle is given by the height of the tree divided by the distance.

Assume the angle θ is 30 degrees:-

$$\text{Tangent } 30° = 0.577 = \frac{\text{height}}{\text{distance}}$$

So if the distance were 100 metres and the angle was 30 degrees, the height of the tree works out to be 57.7 metres.

The same principle is used to determine the distance of the Sun from the Earth. Because the Sun is a vast distance away, approximately 150 million kilometres as mentioned at the start of this chapter, we would need to create a fairly large triangle in order to produce a reasonably sized angle that can be accurately measured. Imagine we measured the angle to the Sun from two places on the Earth's equator. Would the diameter of the Earth provide a suitable baseline?

This arrangement is slightly different to the right angled triangle shown in the previous example, being composed of two back to back right angled triangles, but the trigonometrical relationship between the lengths of the sides and the angles still apply. All the angles in a triangle add up to 180 degrees. We can measure the angles from the horizon to the Sun, at either sides of the equator, and so determine the angle θ, and we know the distance across the equator since it is the diameter of the Earth.

However, things are not that simple as the following diagram shows! The diameter of the Earth at the equator is 12,756 kilometres. This would give an angle between the Earth's horizon and the altitude of the Sun at around 89.9976 degrees. The angle θ subtended at the Sun would therefore only be the difference, a mere 0.004872 degrees, or twice 0.002436 degrees.

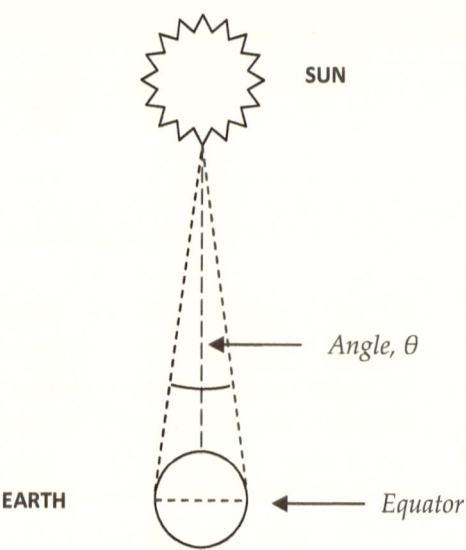

SUN

Angle, θ

EARTH

Equator

Figure 76   USING THE EARTH'S EQUATOR TO TRIANGULATE

The measurements both sides of the world would have to be made simultaneously because the Earth is spinning, rotating 360 degrees a day, 1 degree every four minutes, and the sites on which the measurements are to be taken would need to be very accurately positioned. Although this method could be used in principle, it would be quite difficult to achieve in practice and, as can be seen, the slightest measurement inaccuracy could yield a significant error in the result.

Ideally, what would be needed is to extend the base of the triangle. A baseline far wider than the diameter of the Earth is required. Therefore, to find such a baseline, one needs to look beyond the Earth and out into the Solar System itself.

In the 3rd century BC, Aristarchus, from the Greek island of Samos, had the right idea, and his experiment is referred to as the *Lunar Dichotomy*. He reasoned that when the Moon was in the first or third quarter, when it was at "half Moon", it had to be at right angles to the Sun. The Moon is illuminated by the Sun and at half Moon, it is clearly visible during day. Then he measured the angle between the Sun and the Moon, and from that deduced that the Sun must be between 18 and 20 times further away that the Moon is. Sadly, he did not know how far away the Moon was anyway. His measurements would have yielded a distance to the Sun of only 3.5 to 5 million miles or 5.5 to 8 million kilometres, so it is probably just as well!

The way the measurement is actually made today is as follows. Astronomers make use of the orbit of our nearest planet, Venus. As Venus goes round the Sun, it appears to move to and fro, out to the left and right, at the extremes of its orbit.

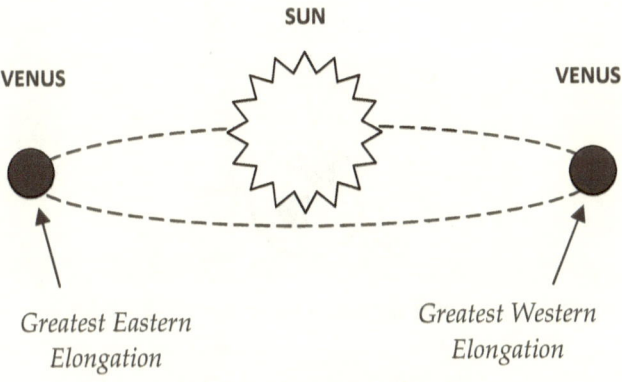

Figure 77   ORBIT OF VENUS

These extremes in the orbital path shown in figure 77 are called the Greatest Western Elongation and Greatest Eastern Elongation respectively. When these greatest elongations are reached, they effectively form a right-angled triangle with the

baseline between the Sun and Venus and the apex here on
Earth.

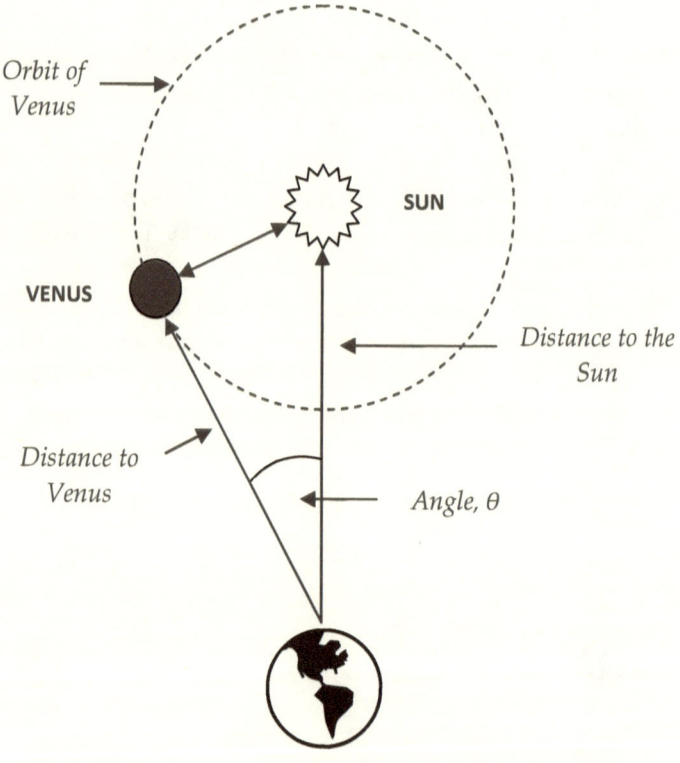

Orbit of Venus

SUN

VENUS

Distance to the Sun

Distance to Venus

Angle, θ

Figure 78   USING THE ORBIT OF VENUS TO TRIANGULATE

We can measure the angle between the Sun and Venus, which
falls within the range 45 to 47 degrees.

Though for simplicity in the diagram, the orbit of Venus is
depicted as being circular, in practice, both the orbits of Venus
and the Earth are slightly elliptical, and this is why there is a
small range in the angles of elongation. To further complicate

matters the orbits of Earth and Venus are not exactly in the same plane. The Earth orbits around the Sun in the plane of the ecliptic, but the orbital plane of Venus is inclined to the plane of the ecliptic, and hence that of the Earth, by 3.39°, as shown below. This also means that the East and West Greatest Elongation distances will differ slightly from one to the other.

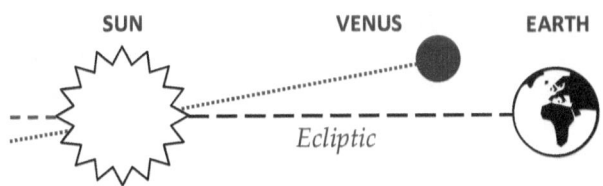

We can accurately measure the distance to Venus using radar! This was first achieved on the 10 May 1961 from a ground based radar. In fact it has been found that Venus is a better radar reflector than the Moon. Since then, using sophisticated radar techniques and probes such as "Magellan", launched in 1989, the surface of Venus has been mapped, recording the depth of valleys and the heights of mountains to a remarkable degree of accuracy. Radar has also been used to accurately measure its rotational parameters.

Amateur radio astronomers have successfully bounced radio signals off Venus using modified television satellite dish aerials, receiving equipment, and transmitters.

Although possible, radar ranging directly on the Sun itself is not practical for a variety of reasons. The Sun is a ball of hot gas and a great deal of the transmitted radar pulse would be absorbed, limiting the intensity of the return echo. Venus is a solid mass with better reflecting properties. In addition, the

Sun has a strong magnetic field, which is also in state of turmoil, and magnetic fields can disrupt radar and radio transmissions. The magnetic field surrounding Venus is lower than that of the Earth.

Water molecules and pollutants in the Earth's atmosphere attenuate all radio and radar signals, but once beyond our atmosphere there is little to attenuate them. Receiving radar echoes from Venus, some 38 million kilometres distance at its nearest approach to us, is therefore perfectly achievable. The return journey of the radar pulse takes around 4 minutes 13 seconds, although at the maximum elongations when it is at right angles to the Sun, the radar distance now extends to around 103 million kilometres, nearly 64 million miles, and the radar echo return time now increases to about 11½ minutes. Radar signals travel at the speed of light, a little less than 300,000 kilometres a second. Therefore, the distance to Venus (in kilometres) is half the time between the transmitted pulse and the return echo, in seconds, multiplied by 300,000.

Once again, the distance of the Earth to the Sun (at the time of the measurement) is given by another formula using another trigonometrical operator called a "cosine". The cosine of the angle θ is the ratio of the distance to Venus divided by the distance to the Sun.

So, $$\text{distance to the Sun} = \frac{\text{distance to Venus}}{\cos \theta}$$

As already stated, like the Earth, the orbit of Venus is not circular but has a degree of ellipticity. The orbital period of Venus is about 224 days, different to that of the Earth at 365 days. The time between reaching the elongations will therefore be around 112 days, which means that from Earth we could observe three such elongations a year. By taking many such observations and measurements over a period of time, the orbital characteristics of both Venus and the Earth can also be determined using these formulae.

The furthest distance we orbit from the Sun, aphelion, is 152.1 million kilometres or 94.53 million miles. The nearest distance, perihelion, is 147.1 million kilometres or 91.42 million miles. The average distance is taken as 149.6 million kilometres or 92.98 million miles. Astronomers use a general unit of distance called the *Astronomical Unit* or AU as being the average distance between the Sun and the Earth.

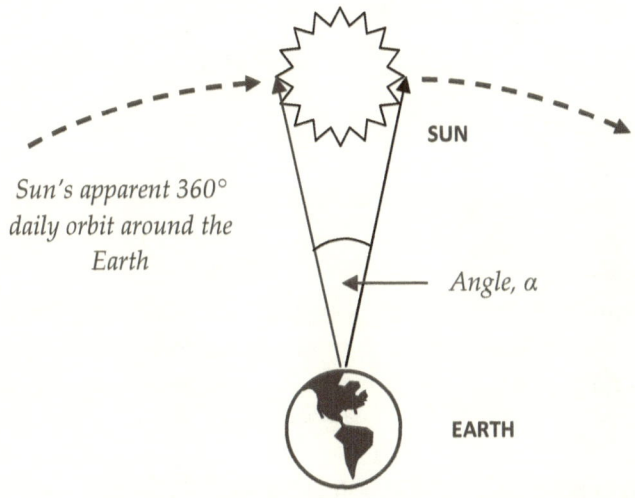

Sun's apparent 360° daily orbit around the Earth

SUN

Angle, α

EARTH

Figure 80   ANGULAR DIAMETER OF THE SUN

We can plot our precise distance from the Sun at any time (during the day) by a simple method; we measure the width, or the diameter of the Sun in degrees. As we are taking spot readings at any one point in time, we can consider the Sun to be effectively orbiting the Earth in a circular fashion because the effect of an elliptical orbit only becomes apparent when viewed over a period of time.

So, let us take as an example a measurement taken on 20 August 2010, when the Greatest Eastern Elongation was 46°. Using radar, the recorded distance to Venus was 105.14 million kilometres. The width of the Sun, or its angular diameter, "α", was also measured on that day to be 31' 40", or expressed as a fraction of a degree becomes 0.52778°. Using the previous equation, we get a distance, r, to the Sun of 151.35 million kilometres. Now "r" represents the <u>radius</u> of a circle (the Sun's apparent celestial orbit) with a circumference of 2πr. With 360° in a circle, we can translate this into a figure whereby the Sun's actual diameter will be:-

$$\text{diameter} = 2\pi \times \frac{0.52778}{360} \times 151.35 \text{ million kilometres}$$

which gives a diameter of 1.394 million kilometres.

The current generally accepted value of the Sun's diameter is 1.392 million kilometres, but it should be remembered that the Sun is in a state of constant turmoil. Its diameter is in a state of flux, and is changing slightly all the time, so the value given above is an average. Knowing the Sun's actual (equatorial) diameter and by measuring the angular diameter, α, we can calculate how far away the Sun is for any day of the year and plot out the precise annual elliptical orbital path that the Earth takes around the Sun. As an example, at perihelion on January 4, the angle is 32' 35"or 0.54306°, and so by rearranging the equation:--

$$\text{distance, d} = \frac{\text{Sun's diameter} \times 360}{2\pi\alpha}$$

which gives us an answer of 147.074 million kilometres, as expected.

By measuring the distance to Venus using radar at the time of its greatest orbital elongations and by simply taking a series of angular measurements, we are able to determine just how far away the Sun is at any time, *and* the precise nature of our own elliptical orbit around the Sun, *and* the Sun's equatorial diameter.

Using this information, the distances and orbital characteristics of all the other bodies in the Solar System can also be determined.

How Big is the Sun?

We have seen in the previous chapter, *"How Far Away is the Sun?"* that the distance to the Sun, though changing slightly due to the fact that our annual orbit is slightly elliptical, is known for any point in time. We have seen how this is achieved by using radar ranging measurements on Venus and how it can accurately determine the Sun's distance.

The Sun's angular diameter, *part of its 360° apparent circular daily orbit around the Earth,* will change on a day to day basis, as shown in figure 80 in the previous chapter. As we approach the Sun it will enlarge and as we recede from the Sun, it will decrease, and all that is necessary to get a day-by-day measure of its distance would be to measure this angle between the opposing outer limbs.

Using simple geometry, the Sun's diameter, or base of the triangle, would be given by twice the distance to the centre of the Sun's disc times the tangent of half the angle:--

$$\text{diameter} = 2 \times \text{distance} \times \text{tangent} \ \frac{\alpha}{2}$$

Using the radar derived data obtained, as previously described as a basis, astronomers can calculate this angle for any point in time. Astronomers publish this angle for every day of the year. It varies from around 32′ 35″ at perihelion, to 31′ 31″ at aphelion.

If this measurement were simply to be made on Earth, then some error is likely to slip in due to distortion of the Sun's image by our atmosphere. When the Sun is low in the sky, at sunset or sunrise for example, the Earth's atmosphere tends to enlarge the image, and layers of air at different temperatures and air turbulence can further distort it. In addition, since the Sun is a gas and not a solid, its diameter is always changing slightly anyway and is never really a perfect sphere.

This measurement could be made by a satellite orbiting outside of Earth's atmosphere, with some adjustment in the formula to account for the position of the satellite. In fact, at the time of writing, there are satellites probing the Sun with specialised equipment to map surface conditions on both the near and the far side.

Given the distance and the angular diameter, the mean diameter of the Sun is calculated to be 1,392,000 kilometres, as a rule, or about 109 times that of the Earth.

The next thing we would want to know is what is the mass of the Sun? This is an easy thing to measure, but first one needs to understand about Isaac Newton's Universal Constant of Gravitation, which will be designated "G". Note, this is not the same thing as the force of gravity, F.

There is a story that Isaac Newton "discovered" gravity when an apple fell from a tree onto his head. In reality, Newton saw an apple fall from a tree, which made him ponder how gravity knows how far up the tree it needs to go, or if there was some limit to how far the effect of gravity extended. If he were to jump off a tall building, he would certainly fall, but the Moon does not fall down onto the Earth.

He knew that a cannonball fired from a cannon will eventually slow down and fall to the ground under the influence of gravity.

Figure 81   TRAJECTORY OF AN OBJECT OVER A FLAT SURFACE

Shown in the next diagram, Isaac Newton postulated that if the cannonball were to be fired with sufficient force, its fall to Earth could follow the curvature of the Earth and it would then proceed to orbit the Earth in the same manner as the Moon does.

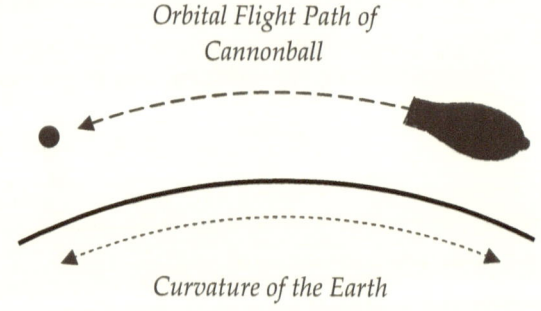

Orbital Flight Path of
Cannonball

Curvature of the Earth

Figure 82   TRAJECTORY OF AN OBJECT OVER A CURVED SURFACE

In 1687, Isaac Newton published his book "Principia Mathematica", in which he declared that all bodies in space are gravitationally attracted to each other, and in relation to a Universal Constant of Gravity. It was not until 1798 when a scientist called Henry Cavendish (who also discovered hydrogen) finally measured this constant. The value of this Constant of Gravity is taken today to be–

$$G = 6.674 \times 10^{-11} \text{ Newton.metre}^2/\text{kilogram}^2$$

Newton's Law of Gravity states that the force of gravity, F, between two bodies is proportional to the sum of their two masses divided by the square of the distance between them, and follows the formula:--

$$F = \frac{G.(\text{mass1} \times \text{mass2})}{\text{distance}^2}$$

Now the force of gravity pulling on an orbiting object is exerting a centripetal force in the same way as swinging a bag of apples around in circles fast enough so they do not all fall out.

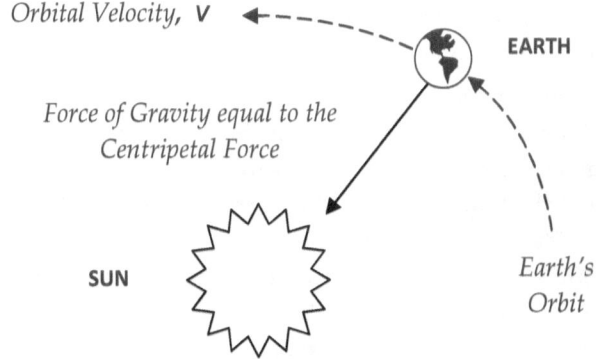

Figure 83   THE EARTH'S ORBITAL PATH DETERMINED BY THE SUN'S GRAVITATIONAL FORCE

Therefore, we can use the orbital parameters of the Earth to determine the mass of the Sun, Ms. Centripetal force is given by the mass of the Earth, Me, times the square of its orbital velocity, v, divided by the mean distance from the Sun, d, and is equal to the force of gravity, F.

$$F = \frac{Me.v^2}{d} = \frac{G.Me.Ms}{d^2}$$

We do not need to know the mass of the Earth since Me is common to both sides of the equation and can therefore be cancelled out. By simply rearranging the formula, we can determine the mass of the Sun, Ms.

$$Ms = \frac{v^2d}{G}$$

We can calculate the Earth's *average* orbital velocity since its orbit is the circumference of a circle described by one rotation a year at an average distance of about 150 million kilometres. From these calculations as a basis, NASA quotes a figure for the mass of the Sun of 1.9891×10^{30} kilogrammes. In round terms this is nearly 2 billion billion billion metric tonnes, and accounts for around 98% of all the mass of the entire Solar System. However, as stated, this assumes that our orbit around the Sun is circular. Our orbit is only slightly elliptical, about 5%, written as 0.05, and so this gives a close estimate of the Sun's mass. A more accurate result can be obtained from studying the orbits of the two inner planets Mercury and Venus, and the next outer planet Mars.

Back in the early 1600's, a German astronomer Johannes Kepler, proposed three laws of planetary motion. He declared that planetary orbits are slightly elliptical and that the radius between the planet and the Sun sweeps out equal areas in equal times. He formulated from his detailed observation of the orbits of the planets around the Sun, that the square of the orbital period, the planet's "year", was proportional to the cube of its *mean distance* from the Sun. This meant that the Sun's mass could be determined from knowing its distance from the Earth and the actual length of the year. The mass of the Sun can also be determined by using the orbital parameters of any of the other planets. Using the "mean distance" assumes circular orbits but since their actual eccentricities are quite low it still provides a reasonable answer, with an accuracy better than 95%. An exact answer can be found by computing the precise planetary orbits and adjusting the equations accordingly. This requires the use of detailed and complex computations which results in the answer quoted by NASA.

Now that we have the mass of the Sun and its diameter, we can calculate its *average* density. Density, or specific gravity, is the ratio of the mass of an object divided by its volume. As we already know the diameter (which is twice the radius), we can calculate the volume.

The volume of a sphere, V, is $\frac{4}{3} \pi r^3$, where "r" is the radius.

This gives a figure for the Sun's volume of 1.412×10^{18} cubic kilometres, and that for the mean density of 1,408 kilogrammes per cubic metre. This amounts to about a quarter of the average density of the Earth, quoted at 5,515 kilogrammes per cubic metre. Although the Sun is composed mainly of hydrogen gas, it is still denser than water (at 1,000 kilogrammes per cubic metre) due to its temperature and its own forces of gravity.

So from some careful original observations, many additional parameters can be calculated, as demonstrated.

How Old is the Sun?

This is a tough one! By reviewing the process of stellar evolution as detailed in the chapter *"The Life Cycle of Stars"*, it is difficult to pick an exact point in time when the Sun was actually born since no precise milestone exists.

Astronomers assess the age of stars by measuring their chemical composition, in particular the ratio of the volume of hydrogen to that of helium. We have seen that as a star ages on the Main Sequence path, it converts hydrogen into helium and so the ratio diminishes with time. This therefore can be used as a guide, but only as a guide.

As a star ages its mass changes and its rate of spin slows down. Without observable surface features, it is virtually impossible to measure spin rate, and it is generally accepted that astronomers are unable to determine the age of single stars. Many stars occur as binary systems, which may contain two, three, four or more interacting bodies. It is assumed that binary stars and star clusters all formed at the same time. Using mathematical modelling based on Jeans' Criteria, an assessment can be made of the time it takes for the gases in the nebula to coalesce and form into these star clusters or binary systems. As such, an estimate of a star's original birth mass can also be made.

Accuracies of around 15% are claimed.

In the case of our Sun, more clues are available, but it must be remembered that a point in time has to be chosen when birth is considered to have taken place. Our Solar System has within it an asteroid belt that occupies for the main part, but not exclusively, the space between the orbits of Mars and Jupiter. The asteroid belt is a region of space populated with millions of rock and metallic fragments. Some of the largest asteroids are called minor planets and are given individual names, such as Vesta, Pallas and Hygiea. These minor planets have diameters or lengths of hundreds of kilometres, and the largest, Ceres, is 950 kilometres in length. Asteroids range in size down to the least, inconsequential, being no larger than a speck of dust.

The asteroid belt was formed at the same time as the Solar System from the same primordial nebula, and is populated with solidified remnants that never evolved into planets or stars. Perturbations in the massive gravitational field of Jupiter accelerated these planetoids in their orbits and prevented them from ever congregating together to form another planet. Indeed, some of the fragments may have originated from a small planet that was broken up by Jupiter's massive gravity. Even today, this is an unstable region of the Solar System and collisions are regularly still taking place.

As already stated, the asteroid belt is not exclusively limited to the region between Mars and Jupiter's orbits, and some material resides inside Mars' orbit between Mars and the Earth. These asteroids are monitored and logged (in part) and are termed "Near Earth Objects" abbreviated as NEO's, or PHA's, "Potentially Hazardous Asteroids". There may be a thousand or more at any one time with the potential of striking the Earth, and they vary in size from a few tens of metres to a few kilometres across.

This is but one source of the thousands of meteors that enter Earth's atmosphere each year. There are others such as the residue of comets but these are usually no bigger than a grain of sand and rarely survive their journey through our atmosphere. With entry speeds between 40,000 and 60,000 miles per hour, most meteors burn up in our atmosphere, but some make it all the way to the surface, and these are called meteorites. Another source is material ejected into space by impacts on the surfaces of Mars and the Moon, which find their way to Earth. Their unique geological features easily distinguish meteorites from native terrestrial rocks.

It is estimated that about 300 tonnes a year actually falls to Earth.

These pieces of space debris are then the same age as the Solar System and therefore the same age as our Sun. So, all that is needed to determine the age of the Sun is to date these meteorites.

This is done by chemical analysis where the presence and relative quantities of particular isotopes are tested. As the name suggests, the half life of a radioactive isotope is the time it takes for the radiation level to decay to half its previous intensity. These meteorites carry uranium isotopes, amongst others, as a legacy from the time of their creation. It is known that uranium 238 decays into lead 206 with a half life of 4.468 billion years. Uranium 235 decays into lead 207 with a half life of 703.8 million years. Uranium isotopes are chosen since they have half lives that span the eons involved and therefore remain in sufficient quantities to facilitate a meaningful measurement. Isotopes with shorter half lives may leave only trace residues making detection and measurement much more difficult and prone to error. The amounts of isotopes that were originally present is unknown, but by comparing the relative amounts of isotopes with differing decay rates, one with another, the age of the meteorite and hence the time elapsed from the creation of our Sun and the Solar System, can be calculated.

Found in Morocco in 2004, the oldest known meteorite tested to date, catalogued as NWA 2364, weighed about 1.5 kilogrammes. (The catalogue alphas "NWA" stand for North West Africa). The age of this meteorite, determined by its isotope composition, is 4.5682 billion years. As such, the age of the Sun may be taken as this figure, and most textbooks quote a figure rounded to 4.6 billion years.

How Much Longer will the Sun Live?

We can only guess how much longer the Sun has to live but to get an idea we need to go back to the chapter on *"The Life Cycle of Stars"*. As with the question of how old is the Sun, we had to decide on exactly at what point in its formation we considered it to have been born. And so it is with the question "at what point in the stellar evolutionary cycle do we consider the Sun to have become extinct". Moreover, the question seeks to expose the secrets of the future and so the answers can only ever be a guess based on observations and calculation. Needless to say, a range of answers can be found in many textbooks and scientific papers, ever changing as new measurements are taken and discoveries made.

However, we do know how much hydrogen and helium the Sun has and we do know the current rate at which the conversion is taking place. We can make a rough calculation therefore, to determine when all the hydrogen will have been consumed, and further nucleosynthesis will cease. The only other information available is to look out into space at the billions of stars, of different ages and masses, and compare them to the condition of our own Sun.

The life cycle of stars is that which is currently believed to be the process. It is the subject of much research and new discoveries are being made all the time. The future course of events that our Sun is likely to experience is believed to be as follows.

It is generally expected that a star like our Sun having lived for 4.5 billion years on the Main Sequence, will continue in its present state for about another 5 to 6 billion years. Some sources claim only 3 to 4 billion years, whereas others claim up to 7 billion years, with 5 to 6 billion years being the current favourite.

As the hydrogen is consumed, its mass will diminish. During this process, it will lose more and more matter drifting off into space as part of the Solar Wind, because its gravity will be decreasing at the same time. This also means that with a lower strength of gravity to contain it, it will start to expand. Some sources claim that in a billion years from now, the Sun will expand about another 10%, sufficient to boil off the Earth's oceans.

This cooling expanding Sun will eventually become a red giant.

Calculations suggest that it will expand to the point where it will engulf Mercury and Venus, but whether it will actually engulf the Earth or not, is uncertain. Another factor associated with the reduction in the Sun's force of gravity is the fact that the Earth will drift further out into space to assume an orbit in the vicinity currently occupied by Mars today.

Some casual estimates made from mathematical models suggest that the Earth will have heated up, due to its proximity to the Sun, to become uninhabitable for about 500,000 years before this point is reached.

As a red giant it will continue to cool for a period thought to be about 100 million years. During this time, the helium converts into higher order elements, up to iron. The core slowly turns into molten iron. As the process continues, a point will be reached when the core collapses altogether under the force of gravity, and the remaining outer gaseous layers will be blown off, to drift away into outer space. This will leave just the very hot molten, and bright luminous core. Our Sun will now become a white dwarf. As a white dwarf, estimates suggest that it will have reduced in size to about that of the Earth!

Thereafter it will continue to cool in the cold vastness of space, becoming forever dimmer until it goes out altogether and vanishes from view. How long this will take is not known.

It will then become a black dwarf, just a cold lump of iron.

Our Sun, the Final Frontier

Astronomers plumb the farthest reaches of the Universe hoping to unravel the mysteries of creation. Of all the countless billions of stars in the heavens there is but one without which we could not exist. It affects each and every one of us in our daily lives, it took part in our creation and will become the architect of our demise.

Throughout this book, we have explored the life cycle of stars, and that of our Sun, as we currently understand it to be. We have peered into the future to try to determine what lies ahead for the Sun, and the destiny of our own tiny planet. We have also unravelled many of the Sun's secrets and discussed the manner by which this was achieved.

There is a plethora of data available in a myriad of textbooks, scientific papers and on the Internet. This data is changing all the time with the advent of new discoveries, and advances in technology, which enable us to make measurements and observations that are ever more accurate.

We have seen how the Sun affects us all in our daily lives. In support of the many ground based solar observatories, a veritable armada of satellites have been deployed by many nations to monitor its behaviour, to try to predict and understand its moods for our own advantage and protection.

This is an exciting time in Astronomy as discoveries are being made at an unprecedented rate. Nevertheless, there is much that we still do not know. For those interested in the subject, solar astronomy is an absorbing and fascinating science, well worth pursuing, that anyone can indulge in from school age on into retirement. Amateur astronomers make a valuable contribution in support of the professionals and your observations could be used to benefit all humanity. Many amateurs contribute their work at a national level, and you could do the same.

There are a billion discoveries out there just waiting to be made by someone, and that someone could well be you!

Appendix 1

Sun Facts

This Appendix tabulates all the solar data discussed in this book

1) Distance between the Earth and the Sun:---

   Maximum 152.1 million kilometres, 94.53 million miles.
   Average 149.6 million kilometres, 92.98 million miles.
   Minimum 147.1 million kilometres, 91.42 million miles.

2) Diameter of the Sun:---

   1.392 million kilometres, 865,000 miles.

3) Mass:---

   1.9891 x 10^{30} kilogrammes, 1.9891 billion billion billion metric tonnes, 1.954 billion billion billion Imperial tons.

4) Volume:---

   1.412 x 10^{18} cubic kilometres, 1.412 billion billion cubic kilometres, 339 million billion cubic miles.

5) Average Density:---

   1,408 kilogrammes per cubic metre, (1.4 times water), 87.86 pounds per cubic foot.

6) Sidereal Rotational Periods (differential rotation):---

Equatorial = 587.36 hours, 14.71° per day.
At latitude 16° = 601.25 hours, 14.37° per day.

7) Solar Constant:---

Total radiated energy 386 billion billion megawatts,
63.41 megawatts per metre^2.

Flux density reaching the Earth' atmosphere:---

Maximum 1,412 Watts per metre^2 at perihelion (January).
Average   1,366 Watts per metre^2.
Minimum 1,321 Watts per metre^2 at aphelion (July).

Flux density measured at ground level (Germany), 800 to
1,100 Watts per metre^2.

8) Regions of the Sun:---

i) The Photosphere.

Average temperature of the surface is 5,776K, 5,500°C.,
ranging from 4,000K to 6,000K, with a thickness of 200 to
500 kilometres, about 125 to 300 miles.

ii) The Convection Zone.

Average temperature unknown, temperature at its
interface with the Radiative Zone estimated to be about
2,000,000K decreasing to around 4,500K to 7,000K at the
base of the Photosphere. It is generally considered to

occupy the outer 30% of the Sun from around 970,000 kilometres from the centre, though some theories suggest it may be much thinner than this.

iii)   The Radiative (Radiation) Zone.

Average temperature believed to be around 5,000,000K decreasing to around 2,000,000K at the interface with the Convection Zone. Temperature at its interface with the solar core estimated to be about 15,700,000K. It is thought to occupy a region from 20 to 25% of the Sun's radius to somewhere about 70% to where it transmutes into the Convection Zone.

iv)   The Core.

This is the nuclear powerhouse of the Sun, where, for the chemical synthesis to take place, temperatures at or above 15.7 million K are required. It is reckoned to occupy the innermost 20 to 25% of the Sun's volume. Various sources give a wide range of estimates for the pressure in the core's centre between 340 million to 225 billion times that of the Earth's atmosphere.

v)   The Chromosphere.

The chromosphere is a rudimentary kind of shallow low pressure atmosphere resident above the photosphere reaching an altitude between 2,500 kilometres and 10,000 kilometres. The temperature just above the photosphere is 4,400K, rising with height to between 30,000K and 100,000K.

vi)   The Corona.

This, again, is an outer region of extremely rarefied gas, much thinner than the chromosphere, which extends out into space for millions of kilometres. Some sources quote temperatures of 1 million K to 2 million K, other sources go as high as 10 million K.

9)   Solar Wind:---

Composed of plasma: protons, electrons and ions. Velocity 200 to 900 kilometres per second, (about ½ to 2 million miles per hour).

Average velocity 468 kilometres per second, (approximately 1 million miles per hour).

The Sun ejects on average 6.7 billion tonnes of material per hour, (just under 2 million tonnes per second).

Average proton count 8.2 million per cubic metre.

10)   Spectral Type::---

Classified on the Hertzprung – Russell chart as class G2v.

Various organisations assign their own slightly different limits to the temperature ranges. "G" is a yellow star and the "2" is part of a *Morgan-Keenan* designation meaning it is two tenths of the range between yellow (6,000K) and orange (5,000 K), or 6,000K minus 2/10 of the 1,000K range, which is. 200K, making it 5,800K, (5,776K). The Roman numeral "v" relates to the width of certain spectral emission lines providing data on a star's size and luminosity.

11) Sunspot Umbral Field Strengths

Major sunspots 1,000 to 4,000 Gauss, (0.2 to 0.4 T).
Small sunspots as low as 10 Gauss, (10^{-3} T).

12) Age of the Sun:---

4.6 billion years.

13) Life Expectancy:---

The general consensus is that it will continue for another 5 to 6 billion years to eventually become a red giant. Various sources range from 3 to 7 billion years.

At that point it will collapse under its own gravity over a further period of about 100 million years to become a white dwarf.

Thereafter it will cool and go out, finally becoming a black dwarf over an indeterminate period.

Some of these figures may well change with time as measuring techniques and equipment advance.

Appendix 2

Chemical Composition of the Sun

The first table shows how the chemical composition is presented, comparing the abundance of a chemical element's atoms in relation to one trillion hydrogen atoms. Shown alongside are the logarithmic values. All 92 elements are listed, even where no trace has been detected.

Table 1

Atomic Number	Name	$Log_{10}H$	Number of Atoms
1	HYDROGEN	12.00	1000000000000.00
2	HELIUM	10.93	85113803820.24
3	LITHIUM	1.05	11.22
4	BERYLLIUM	1.38	23.99
5	BORON	2.70	501.19
6	CARBON	8.43	269153480.39
7	NITROGEN	7.83	67608297.54
8	OXYGEN	8.69	489778819.37
9	FLOURINE	4.56	36307.81
10	NEON	7.93	85113803.82
11	SODIUM	6.24	1737800.83
12	MAGNESIUM	7.60	39810717.06
13	ALUMINIUM	6.45	2818382.93
14	SILICON	7.51	32359365.69
15	PHOSPHORUS	5.41	257039.58
16	SULPHUR	7.12	13182567.39
17	CHLORINE	5.50	316227.77

Table 1 continued

Atomic Number	Name	$Log_{10}H$	Number of Atoms
18	ARGON	6.40	2511886.43
19	POTASSIUM	5.03	107151.93
20	CALCIUM	6.34	2187761.62
21	SCANDIUM	3.15	1412.54
22	TITANIUM	4.95	89125.09
23	VANADIUM	3.93	8511.38
24	CHROMIUM	5.64	436515.83
25	MANGANESE	5.43	269153.48
26	IRON	7.50	31622776.60
27	COBALT	4.99	97723.72
28	NICKEL	6.22	1659586.91
29	COPPER	4.19	15488.17
30	ZINC	4.56	36307.81
31	GALLIUM	3.04	1096.48
32	GERMANIUM	3.65	4466.84
33	ARSENIC	0	0
34	SELENIUM	0	0
35	BROMINE	0	0
36	KRYPTON	3.25	1778.28
37	RUBIDIUM	2.52	331.13
38	STRONTIUM	2.87	741.31
39	YTTRIUM	2.21	162.18
40	ZIRCONIUM	2.58	380.19
41	NIOBIUM	1.46	28.84
42	MOLYBDENUM	1.88	75.86
43	TECHNETIUM	0	0
44	RUTHENIUM	1.75	56.23
45	RHODIUM	0.91	8.13

Table 1 continued

Atomic Number	Name	Log₁₀H	Number of Atoms
46	PALLADIUM	1.57	37.15
47	SILVER	0.94	8.71
48	CADMIUM	0	0
49	INDIUM	0.80	6.31
50	TIN	2.04	109.65
51	ANTIMONY	0	0
52	TELLURIUM	0	0
53	IODINE	0	0
54	XENON	2.24	173.78
55	CAESIUM	0	0
56	BARIUM	2.18	151.36
57	LANTHANUM	1.10	12.59
58	CERIUM	1.58	38.02
59	PRASEODIMIUM	0.72	5.25
60	NEODYMIUM	1.42	26.30
61	PROMETHIUM	0	0
62	SAMARIUM	0.96	9.12
63	EUROPIUM	0.52	3.31
64	GADOLINIUM	1.07	11.75
65	TERBIUM	0.30	2.00
66	DYSPROSIUM	1.10	12.59
67	HOLMIUM	0.48	3.02
68	ERBIUM	0.92	8.32
69	THULIUM	0.10	1.26
70	YTTERBIUM	0.84	6.92
71	LUTETIUM	0.10	1.26
72	HAFNIUM	0.85	7.08
73	TANTALUM	0	0

Table 1 continued

Atomic Number	Name	$Log_{10}H$	Number of Atoms
74	TUNGSTEN	0.85	7.08
75	RHENIUM	0	0
76	OSMIUM	1.40	25.12
77	IRIDIUM	1.38	23.99
78	PLATINUM	0	0
79	GOLD	0.92	8.32
80	MERCURY	0	0
81	THALLIUM	0.90	7.94
82	LEAD	1.75	56.23
83	BISMUTH	0	0
84	POLONIUM	0	0
85	ASTATINE	0	0
86	RADON	0	0
87	FRANCIUM	0	0
88	RADIUM	0	0
89	ACTINIUM	0	0
90	THORIUM	0.02	1.05
91	PROTACTINIUM	0	0
92	URANIUM	0	0

The second table lists the percentage volume of each element. These percentage volume figures are obtained by dividing the number of atoms for any particular chemical element into the sum total of all the atoms in the list, multiplied by 100:---

$$\% \ volume = \frac{Number \ of \ atoms \ for \ that \ particular \ element \times 100}{Total \ count \ of \ all \ the \ atoms \ in \ the \ table}$$

Table 2

Atomic Number	Atomic Symbol	Name	% volume
1	H	HYDROGEN	92.067888280846
2	He	HELIUM	7.836248181279
3	Li	LITHIUM	0.000000001033
4	Be	BERYLLIUM	0.000000002209
5	B	BORON	0.000000046143
6	C	CARBON	0.024780392563
7	N	NITROGEN	0.006224553185
8	O	OXYGEN	0.045092901624
9	F	FLOURINE	0.000003342783
10	Ne	NEON	0.007836248181
11	Na	SODIUM	0.000159995653
12	Mg	MAGNESIUM	0.003665288650
13	Al	ALUMINIUM	0.000259482565
14	Si	SILICON	0.002979258465
15	P	PHOSPHORUS	0.000023665091
16	S	SULPHUR	0.001213691141
17	Cl	CHLORINE	0.000029114423
18	Ar	ARGON	0.000231264079
19	K	POTASSIUM	0.000009865252
20	Ca	CALCIUM	0.000201422593
21	Sc	SCANDIUM	0.000000130049
22	Ti	TITANIUM	0.000008205559
23	V	VANADIUM	0.000000783625
24	Cr	CHROMIUM	0.000040189091
25	Mn	MANGANESE	0.000024780393
26	Fe	IRON	0.002911442263
27	Co	COBALT	0.000008997217
28	Ni	NICKEL	0.000152794662

Table 2 continued

Atomic Number	Atomic Symbol	Name	% volume
29	Cu	COPPER	0.000001425963
30	Zn	ZINC	0.000003342783
31	Ga	GALLIUM	0.000000100950
32	Ge	GERMANIUM	0.000000411252
33	As	ARSENIC	0
34	Se	SELENIUM	0
35	Br	BROMINE	0
36	Kr	KRYPTON	0.000000163722
37	Rb	RUBIDIUM	0.000000030487
38	Sr	STRONTIUM	0.000000068251
39	Y	YTTRIUM	0.000000014932
40	Zr	ZIRCONIUM	0.000000035003
41	Nb	NIOBIUM	0.000000002655
42	Mo	MOLYBDENUM	0.000000006984
43	Tc	TECHNETIUM	0
44	Ru	RUTHENIUM	0.000000005177
45	Rh	RHODIUM	0.000000000748
46	Pd	PALLADIUM	0.000000003421
47	Ag	SILVER	0.000000000802
48	Cd	CADMIUM	0
49	In	INDIUM	0.000000000581
50	Sn	TIN	0.000000010095
51	Sb	ANTIMONY	0
52	Te	TELLURIUM	0
53	I	IODINE	0
54	Xe	XENON	0.000000016000
55	Cs	CAESIUM	0
56	Ba	BARIUM	0.000000013935

Table 2 continued

Atomic Number	Atomic Symbol	Name	% volume
57	La	LANTHANUM	0.000000001159
58	Ce	CERIUM	0.000000003500
59	Pr	PRASEODIMIUM	0.000000000483
60	Nd	NEODYMIUM	0.000000002422
61	Pm	PROMETHIUM	0
62	Sm	SAMARIUM	0.000000000840
63	Eu	EUROPIUM	0.000000000305
64	Gd	GADOLINIUM	0.000000001082
65	Tb	TERBIUM	0.000000000184
66	Dy	DYSPROSIUM	0.000000001159
67	Ho	HOLMIUM	0.000000000278
68	Er	ERBIUM	0.000000000766
69	Tm	THULIUM	0.000000000116
70	Yb	YTTERBIUM	0.000000000637
71	Lu	LUTETIUM	0.000000000116
72	Hf	HAFNIUM	0.000000000652
73	Ta	TANTALUM	0
74	W	TUNGSTEN	0.000000000652
75	Re	RHENIUM	0
76	Os	OSMIUM	0.000000002313
77	Ir	IRIDIUM	0.000000002209
78	Pt	PLATINUM	0
79	Au	GOLD	0.000000000766
80	Hg	MERCURY	0
81	Tl	THALLIUM	0.000000000731
82	Pb	LEAD	0.000000005177
83	Bi	BISMUTH	0
84	Po	POLONIUM	0

Table 2 continued

Atomic Number	Atomic Symbol	Name	% volume
85	At	ASTATINE	0
86	Rn	RADON	0
87	Fr	FRANCIUM	0
88	Ra	RADIUM	0
89	Ac	ACTINIUM	0
90	Th	THORIUM	0.000000000096
91	Pa	PROTACTINIUM	0
92	U	URANIUM	0

The third table shows what percentage of the Sun's mass is attributable to each chemical element. Once again, the list shows all the 92 naturally occurring elements. To obtain these figures the number of atoms of a particular element is multiplied by its atomic weight, and then divided by the sum total of all of the atoms of every element multiplied by their individual atomic weights, then multiplied by 100 to convert it into a percentage:---

$$\% \text{ mass} = \frac{\textit{Number of atoms} \times \textit{atomic weight} \times 100}{\textit{Sum of all (atoms} \times \textit{atomic weights)}}$$

Table 3

Atomic Symbol	Name	Atomic Weight	% mass
H	HYDROGEN	1.008	73.7387795725
He	HELIUM	4.003	24.9241871408

Table 3 continued

Atomic Symbol	Name	Atomic Weight	% mass
Li	LITHIUM	6.941	0.0000000057
Be	BERYLLIUM	9.012	0.0000000158
B	BORON	10.81	0.0000003963
C	CARBON	12.01	0.2364712901
N	NITROGEN	14.01	0.0692904766
O	OXYGEN	16.00	0.5732649587
F	FLOURINE	19.00	0.0000504649
Ne	NEON	20.18	0.1256482879
Na	SODIUM	22.99	0.0029226337
Mg	MAGNESIUM	24.31	0.0707978994
Al	ALUMINIUM	26.98	0.0055625939
Si	SILICON	28.09	0.0664947186
P	PHOSPHORUS	30.97	0.0005823400
S	SULPHUR	32.07	0.0309267564
Cl	CHLORINE	35.45	0.0008200714
Ar	ARGON	39.95	0.0073409488
K	POTASSIUM	39.10	0.0003064871
Ca	CALCIUM	40.08	0.0064145047
Sc	SCANDIUM	44.96	0.0000046458
Ti	TITANIUM	47.88	0.0003121688
V	VANADIUM	50.94	0.0000317172
Cr	CHROMIUM	52.00	0.0016604995
Mn	MANGANESE	54.94	0.0010817429
Fe	IRON	55.85	0.1291988330
Co	COBALT	58.47	0.0004179925
Ni	NICKEL	58.69	0.0071252404
Cu	COPPER	63.55	0.0000720031
Zn	ZINC	65.39	0.0001736788

Table 3 continued

Atomic Symbol	Name	Atomic Weight	% mass
Ga	GALLIUM	69.72	0.0000055923
Ge	GERMANIUM	72.59	0.0000237199
As	ARSENIC	74.92	0
Se	SELENIUM	78.96	0
Br	BROMINE	79.90	0
Kr	KRYPTON	83.80	0.0000109013
Rb	RUBIDIUM	85.47	0.0000020704
Sr	STRONTIUM	87.62	0.0000047516
Y	YTTRIUM	88.91	0.0000010548
Zr	ZIRCONIUM	91.22	0.0000025370
Nb	NIOBIUM	92.91	0.0000001960
Mo	MOLYBDENUM	95.94	0.0000005324
Tc	TECHNETIUM	98.00	0
Ru	RUTHENIUM	101.10	0.0000004159
Rh	RHODIUM	102.90	0.0000000612
Pd	PALLADIUM	106.40	0.0000002892
Ag	SILVER	107.91	0.0000000688
Cd	CADMIUM	112.41	0
In	INDIUM	114.80	0.0000000530
Sn	TIN	118.70	0.0000009522
Sb	ANTIMONY	121.80	0
Te	TELLURIUM	127.60	0
I	IODINE	126.90	0
Xe	XENON	131.30	0.0000016692
Cs	CAESIUM	132.90	0
Ba	BARIUM	137.30	0.0000015202
La	LANTHANUM	138.90	0.0000001279
Ce	CERIUM	140.10	0.0000003896

Table 3 continued

Atomic Symbol	Name	Atomic Weight	% mass
Pr	PRASEODIMIUM	140.90	0.0000000541
Nd	NEODYMIUM	144.20	0.0000002775
Pm	PROMETHIUM	147.00	0
Sm	SAMARIUM	150.40	0.0000000981
Eu	EUROPIUM	152.00	0.0000000368
Gd	GADOLINIUM	157.30	0.0000001352
Tb	TERBIUM	158.90	0.0000000232
Dy	DYSPROSIUM	162.50	0.0000001497
Ho	HOLMIUM	164.90	0.0000000364
Er	ERBIUM	167.30	0.0000001018
Tm	THULIUM	168.90	0.0000000156
Yb	YTTERBIUM	173.00	0.0000000876
Lu	LUTETIUM	175.00	0.0000000161
Hf	HAFNIUM	178.50	0.0000000924
Ta	TANTALUM	180.90	0
W	TUNGSTEN	183.90	0.0000000952
Re	RHENIUM	186.20	0
Os	OSMIUM	190.20	0.0000003495
Ir	IRIDIUM	190.20	0.0000003338
Pt	PLATINUM	195.10	0
Au	GOLD	197.00	0.0000001199
Hg	MERCURY	200.50	0
Tl	THALLIUM	204.40	0.0000001188
Pb	LEAD	207.20	0.0000008524
Bi	BISMUTH	209.00	0
Po	POLONIUM	210.00	0
At	ASTATINE	210.00	0
Rn	RADON	222.00	0

Table 3 continued

Atomic Symbol	Name	Atomic Weight	% mass
Fr	FRANCIUM	223.00	0
Ra	RADIUM	226.00	0
Ac	ACTINIUM	227.00	0
Th	THORIUM	232.00	0.0000000178
Pa	PROTACTINIUM	231.00	0
U	URANIUM	238.00	0

Knowing the overall mass of the Sun, and with the percentage masses of the various chemical elements, their actual masses can be calculated. The masses given in the following table are billion trillion metric tonnes, or 10^{21} metric tonnes.

Table 4

Atomic Symbol	Name	Mass x 10^{21} metric tonnes
H	HYDROGEN	1,466,738.064476
He	HELIUM	495,767.006418
Li	LITHIUM	0.000113
Be	BERYLLIUM	0.000315
B	BORON	0.007883
C	CARBON	4,703.650432
N	NITROGEN	1,378.256870
O	OXYGEN	11,402.813294
F	FLOURINE	1.003796
Ne	NEON	2,499.270095
Na	SODIUM	58.134107

Table 4 continued

Atomic Symbol	Name	Mass x 10^{21} metric tonnes
Mg	MAGNESIUM	1,408.241017
Al	ALUMINIUM	110.645556
Si	SILICON	1,322.646448
P	PHOSPHORUS	11.583325
S	SULPHUR	615.164111
Cl	CHLORINE	16.312040
Ar	ARGON	146.018813
K	POTASSIUM	6.096334
Ca	CALCIUM	127.590913
Sc	SCANDIUM	0.092410
Ti	TITANIUM	6.209350
V	VANADIUM	0.630886
Cr	CHROMIUM	33.028996
Mn	MANGANESE	21.516949
Fe	IRON	2,569.893987
Co	COBALT	8.314289
Ni	NICKEL	141.728157
Cu	COPPER	1.432213
Zn	ZINC	3.454645
Ga	GALLIUM	0.111237
Ge	GERMANIUM	0.471812
As	ARSENIC	0
Se	SELENIUM	0
Br	BROMINE	0
Kr	KRYPTON	0.216838
Rb	RUBIDIUM	0.041182
Sr	STRONTIUM	0.094514
Y	YTTRIUM	0.020982

Table 4 continued

Atomic Symbol	Name	Mass x 10^{21} metric tonnes
Zr	ZIRCONIUM	0.050464
Nb	NIOBIUM	0.003899
Mo	MOLYBDENUM	0.010590
Tc	TECHNETIUM	0
Ru	RUTHENIUM	0.008273
Rh	RHODIUM	0.001217
Pd	PALLADIUM	0.005752
Ag	SILVER	0.001368
Cd	CADMIUM	0
In	INDIUM	0.001054
Sn	TIN	0.018940
Sb	ANTIMONY	0
Te	TELLURIUM	0
I	IODINE	0
Xe	XENON	0.033201
Cs	CAESIUM	0
Ba	BARIUM	0.030239
La	LANTHANUM	0.002544
Ce	CERIUM	0.007751
Pr	PRASEODIMIUM	0.001076
Nd	NEODYMIUM	0.005519
Pm	PROMETHIUM	0
Sm	SAMARIUM	0.001951
Eu	EUROPIUM	0.000732
Gd	GADOLINIUM	0.002689
Tb	TERBIUM	0.000461
Dy	DYSPROSIUM	0.002977
Ho	HOLMIUM	0.000725

Table 4 continued

Atomic Symbol	Name	Mass x 10^{21} metric tonnes
Er	ERBIUM	0.002025
Tm	THULIUM	0.000309
Yb	YTTERBIUM	0.001742
Lu	LUTETIUM	0.000321
Hf	HAFNIUM	0.001839
Ta	TANTALUM	0
W	TUNGSTEN	0.001894
Re	RHENIUM	0
Os	OSMIUM	0.006952
Ir	IRIDIUM	0.006639
Pt	PLATINUM	0
Au	GOLD	0.002384
Hg	MERCURY	0
Tl	THALLIUM	0.002363
Pb	LEAD	0.016954
Bi	BISMUTH	0
Po	POLONIUM	0
At	ASTATINE	0
Rn	RADON	0
Fr	FRANCIUM	0
Ra	RADIUM	0
Ac	ACTINIUM	0
Th	THORIUM	0.000353
Pa	PROTACTINIUM	0
U	URANIUM	0

Appendix 3

Equation of Time Correction Factors

The following table is provided for those readers who may require it. It lists the daily correction factors and is an average taken over a four year period to include Leap Years. The chapter *"The Equation of Time"* details how these figures are used. The *Value* given here is in minutes and seconds.

YEAR DAY	DATE	VALUE	YEAR DAY	DATE	VALUE
1	1 Jan	-03:12	19	19 Jan	-10:32
2	2 Jan	-03:40	20	20 Jan	-10:50
3	3 Jan	-04:08	21	21 Jan	-11:08
4	4 Jan	-04:36	22	22 Jan	-11:25
5	5 Jan	-05:03	23	23 Jan	-11:41
6	6 Jan	-05:30	24	24 Jan	-11:57
7	7 Jan	-05:57	25	25 Jan	-12:12
8	8 Jan	-06:23	26	26 Jan	-12:26
9	9 Jan	-06:49	27	27 Jan	-12:39
10	10 Jan	-07:14	28	28 Jan	-12:51
11	11 Jan	-07:38	29	29 Jan	-13:03
12	12 Jan	-08:02	30	30 Jan	-13:14
13	13 Jan	-08:25	31	31 Jan	-13:24
14	14 Jan	-08:48			
15	15 Jan	-09:10	32	1 Feb	-13:33
16	16 Jan	-09:32	33	2 Feb	-13:41
17	17 Jan	-09:52	34	3 Feb	-13:48
18	18 Jan	-10:12	35	4 Feb	-13:55

YEAR DAY	DATE	VALUE
36	5 Feb	-14:01
37	6 Feb	-14:06
38	7 Feb	-14:10
39	8 Feb	-14:14
40	9 Feb	-14:16
41	10 Feb	-14:18
42	11 Feb	-14:19
43	12 Feb	-14:20
44	13 Feb	-14:19
45	14 Feb	-14:18
46	15 Feb	-14:16
47	16 Feb	-14:13
48	17 Feb	-14:10
49	18 Feb	-14:06
50	19 Feb	-14:01
51	20 Feb	-13:55
52	21 Feb	-13:49
53	22 Feb	-13:42
54	23 Feb	-13:35
55	24 Feb	-13:27
56	25 Feb	-13:18
57	26 Feb	-13:09
58	27 Feb	-12:59
59	28 Feb	-12:48
60	*29 Feb*	*-12:42*
61	1 Mar	-12:34
62	2 Mar	-12:23
63	3 Mar	-12:11

YEAR DAY	DATE	VALUE
64	4 Mar	-11:58
65	5 Mar	-11:45
66	6 Mar	-11:31
67	7 Mar	-11:17
68	8 Mar	-11:03
69	9 Mar	-10:48
70	10 Mar	-10:33
71	11 Mar	-10:18
72	12 Mar	-10:02
73	13 Mar	-09:46
74	14 Mar	-09:30
75	15 Mar	-09:13
76	16 Mar	-08:56
77	17 Mar	-08:39
78	18 Mar	-08:22
79	19 Mar	-08:04
80	20 Mar	-07:46
81	21 Mar	-07:28
82	22 Mar	-07:10
83	23 Mar	-06:52
84	24 Mar	-06:34
85	25 Mar	-06:16
86	26 Mar	-05:58
87	27 Mar	-05:40
88	28 Mar	-05:21
89	29 Mar	-05:02
90	30 Mar	-04:44
91	31 Mar	-04:26

YEAR DAY	DATE	VALUE	YEAR DAY	DATE	VALUE
92	1 Apr	-04:08	121	30 Apr	+02:43
93	2 Apr	-03:50			
94	3 Apr	-03:32	122	1 May	+02:51
95	4 Apr	-03:14	123	2 May	+02:59
96	5 Apr	-02:57	124	3 May	+03:06
97	6 Apr	-02:40	125	4 May	+03:12
98	7 Apr	-02:23	126	5 May	+03:18
99	8 Apr	-02:06	127	6 May	+03:23
100	9 Apr	-01:49	128	7 May	+03:27
101	10 Apr	-01:32	129	8 May	+03:31
102	11 Apr	-01:16	130	9 May	+03:35
103	12 Apr	-01:00	131	10 May	+03:38
104	13 Apr	-00:44	132	11 May	+03:40
105	14 Apr	-00:29	133	12 May	+03:42
106	15 Apr	-00:14	134	13 May	+03:44
107	16 Apr	+00:01	135	14 May	+03:44
108	17 Apr	+00:15	136	15 May	+03:44
109	18 Apr	+00:29	137	16 May	+03:44
110	19 Apr	+00:43	138	17 May	+03:43
111	20 Apr	+00:56	139	18 May	+03:41
112	21 Apr	+01:00	140	19 May	+03:39
113	22 Apr	+01:21	141	20 May	+03:37
114	23 Apr	+01:33	142	21 May	+03:34
115	24 Apr	+01:45	143	22 May	+03:30
116	25 Apr	+01:56	144	23 May	+03:24
117	26 Apr	+02:06	145	24 May	+03:21
118	27 Apr	+02:16	146	25 May	+03:16
119	28 Apr	+02:26	147	26 May	+03:10
120	29 Apr	+02:35	148	27 May	+03:03

YEAR DAY	DATE	VALUE
149	28 May	+02:56
150	29 May	+02:49
151	30 May	+02:41
152	31 May	+02:33
153	1 Jun	+02:25
154	2 Jun	+02:16
155	3 Jun	+02:06
156	4 Jun	+01:56
157	5 Jun	+01:46
158	6 Jun	+01:36
159	7 Jun	+01:25
160	8 Jun	+01:14
161	9 Jun	+01:03
162	10 Jun	+00:51
163	11 Jun	+00:39
164	12 Jun	+00:27
165	13 Jun	+00:15
166	14 Jun	+00:03
167	15 Jun	-00:10
168	16 Jun	-00:23
169	17 Jun	-00:36
170	18 Jun	-00:49
171	19 Jun	-01:02
172	20 Jun	-01:15
173	21 Jun	-01:28
174	22 Jun	-01:41
175	23 Jun	-01:54
176	24 Jun	-02:07

YEAR DAY	DATE	VALUE
177	25 Jun	-02:20
178	26 Jun	-02:33
179	27 Jun	-02:45
180	28 Jun	-02:57
181	29 Jun	-03:09
182	30 Jun	-03:21
183	1 Jul	-03:33
184	2 Jul	-03:45
185	3 Jul	-03:57
186	4 Jul	-04:08
187	5 Jul	-04:19
188	6 Jul	-04:29
189	7 Jul	-04:39
190	8 Jul	-04:49
191	9 Jul	-04:58
192	10 Jul	-05:07
193	11 Jul	-05:16
194	12 Jul	-05:24
195	13 Jul	-05:32
196	14 Jul	-05:39
197	15 Jul	-05:46
198	16 Jul	-05:52
199	17 Jul	-05:58
200	18 Jul	-06:03
201	19 Jul	-06:08
202	20 Jul	-06:12
203	21 Jul	-06:15
204	22 Jul	-06:18

YEAR DAY	DATE	VALUE
205	23 Jul	-06:20
206	24 Jul	-06:22
207	25 Jul	-06:24
208	26 Jul	-06:25
209	27 Jul	-06:25
210	28 Jul	-06:24
211	29 Jul	-06:23
212	30 Jul	-06:21
213	31 Jul	-06:19
214	1 Aug	-06:16
215	2 Aug	-06:16
216	3 Aug	-06:09
217	4 Aug	-06:04
218	5 Aug	-05:59
219	6 Aug	-05:53
220	7 Aug	-05:46
221	8 Aug	-05:39
222	9 Aug	-05:31
223	10 Aug	-05:23
224	11 Aug	-05:14
225	12 Aug	-05:05
226	13 Aug	-04:55
227	14 Aug	-04:44
228	15 Aug	-04:33
229	16 Aug	-04:21
230	17 Aug	-04:09
231	18 Aug	-03:57
232	19 Aug	-03:44

YEAR DAY	DATE	VALUE
233	20 Aug	-03:30
234	21 Aug	-03:16
235	22 Aug	-03:01
236	23 Aug	-02:46
237	24 Aug	-02:30
238	25 Aug	-02:14
239	26 Aug	-01:58
240	27 Aug	-01:41
241	28 Aug	-01:24
242	29 Aug	-01:07
243	30 Aug	-00:49
244	31 Aug	-00:31
245	1 Sep	-00:12
246	2 Sep	+00:07
247	3 Sep	+00:26
248	4 Sep	+00:45
249	5 Sep	+01:05
250	6 Sep	+01:25
251	7 Sep	+01:45
252	8 Sep	+02:05
253	9 Sep	+02:26
254	10 Sep	+02:47
255	11 Sep	+03:08
256	12 Sep	+03:29
257	13 Sep	+03:50
258	14 Sep	+04:11
259	15 Sep	+04:32
260	16 Sep	+04:53

YEAR DAY	DATE	VALUE
261	17 Sep	+05:14
262	18 Sep	+05:35
263	19 Sep	+05:56
264	20 Sep	+06:18
265	21 Sep	+06:40
266	22 Sep	+07:01
267	23 Sep	+07:22
268	24 Sep	+07:43
269	25 Sep	+08:04
270	26 Sep	+08:25
271	27 Sep	+08:46
272	28 Sep	+09:06
273	29 Sep	+09:26
274	30 Sep	+09:46
275	1 Oct	+10:05
276	2 Oct	+10:24
277	3 Oct	+10:43
278	4 Oct	+11:02
279	5 Oct	+11:20
280	6 Oct	+11:38
281	7 Oct	+11:56
282	8 Oct	+12:13
283	9 Oct	+12:30
284	10 Oct	+12:46
285	11 Oct	+13:02
286	12 Oct	+13:18
287	13 Oct	+13:33
288	14 Oct	+13:47

YEAR DAY	DATE	VALUE
289	15 Oct	+14:01
290	16 Oct	+14:14
291	17 Oct	+14:27
292	18 Oct	+14:39
293	19 Oct	+14:51
294	20 Oct	+15:02
295	21 Oct	+15:12
296	22 Oct	+15:22
297	23 Oct	+15:31
298	24 Oct	+15:40
299	25 Oct	+15:54
300	26 Oct	+15:57
301	27 Oct	+16:01
302	28 Oct	+16:06
303	29 Oct	+16:11
304	30 Oct	+16:15
305	31 Oct	+16:18
306	1 Nov	+16:20
307	2 Nov	+16:22
308	3 Nov	+16:23
309	4 Nov	+16:23
310	5 Nov	+16:22
311	6 Nov	+16:20
312	7 Nov	+16:18
313	8 Nov	+16:15
314	9 Nov	+16:11
315	10 Nov	+16:06
316	11 Nov	+16:00

YEAR DAY	DATE	VALUE	YEAR DAY	DATE	VALUE
317	12 Nov	+15:53	342	7 Dec	+08:48
318	13 Nov	+15:49	343	8 Dec	+08:22
319	14 Nov	+15:37	344	9 Dec	+07:56
320	15 Nov	+15:28	345	10 Dec	+07:29
321	16 Nov	+15:18	346	11 Dec	+07:02
322	17 Nov	+15:07	347	12 Dec	+06:34
323	18 Nov	+14:56	348	13 Dec	+06:06
324	19 Nov	+14:43	349	14 Dec	+05:38
325	20 Nov	+14:30	350	15 Dec	+05:09
326	21 Nov	+14:16	351	16 Dec	+04:40
327	22 Nov	+14:01	352	17 Dec	+04:11
328	23 Nov	+13:45	353	18 Dec	+03:42
329	24 Nov	+13:28	354	19 Dec	+03:13
330	25 Nov	+13:11	355	20 Dec	+02:43
331	26 Nov	+12:53	356	21 Dec	+02:13
332	27 Nov	+12:34	357	22 Dec	+01:43
333	28 Nov	+12:14	358	23 Dec	+01:13
334	29 Nov	+11:54	359	24 Dec	+00:43
335	30 Nov	+11:33	360	25 Dec	+00:13
			361	26 Dec	-00:17
336	1 Dec	+11:11	362	27 Dec	-00:47
337	2 Dec	+10:49	363	28 Dec	-01:16
338	3 Dec	+10:26	364	29 Dec	-01:45
339	4 Dec	+10:02	365	30 Dec	-02:18
340	5 Dec	+09:38	366	31 Dec	-02:43
341	6 Dec	+09:13			

Glossary of Terms

Aphelion. That point on Earth's elliptical orbit, at its farthest distance from the Sun.

Black Dwarf. The stellar corpse of a former white dwarf that has cooled and no longer emits heat or light. Black dwarfs are theoretical bodies since the time required for this condition to be reached exceeds the current age of the Universe.

Black Hole. A region surrounding the collapsed stellar core of a once massive star with a extremely high gravitational field from which nothing, not even light, can escape.

Bohr Model. Classical model of the basic atom, comprising a positively charged nucleus composed of bound protons and neutrons, surrounded by orbiting electrostatically attracted electrons. Introduced in 1913 and named after Niels Bohr, Danish physicist. Refer also to "ion" and "isotope".

Boltzmann's Constant. This is a physical constant describing the relationship between the kinetic energy and temperature of a gas and takes the value 1.3806×10^{-23} Joules per Kelvin.

Coulomb. A unit of measure of electric charge, transporting a current of 1 Ampére per second.

Declination. In solar astronomy, it is the angle between the plane of the ecliptic and the Earth's equator. In general astronomy, it is the angle between any celestial body and the Earth's equator. North of the equator the angle is positive, and is negative south of the equator. Refer to diagrams 21 and 22.

Doppler Shift. A Doppler shift is a change in wavelength of either an acoustic or an electromagnetic source occasioned by movement of the source. It is often described as the drop in the pitch of a train whistle, as the approaching train passes the listener. An approaching source will raise the frequency, whereas a receding source will reduce it.

Red shifted stars are those moving away from us where their visible emission spectra have been shifted towards the longer wavelength red end of the spectrum, by an amount proportional to their velocity. *Blue shifted* stars are those moving toward us, where their spectra have been proportionally shifted towards the blue, shorter wavelengths.

Ecliptic. The plane of the ecliptic is the two dimensional plane described by the Earth in its annual orbit around the Sun.

Electron. A negatively charged sub atomic elementary particle, with a charge of 1.602×10^{-19} Coulombs, and a rest mass of 9.11×10^{-31} kilogrammes.

Equinox. Occurring twice annually on March 20 or 21 and September 22 or 23, a condition where the Earth's equator aligns to the plane of the ecliptic, and the Earth's pole of rotation is vertical. Meaning "equal day, equal night". A mean solar day with 12 hours each of both daylight and darkness (approximately).

ESA. European Space Agency. An 18 nation European intergovernmental agency dedicated to space exploration.

Gauss. The CGS unit of measure of magnetic flux density, denoted by the letter "G". CGS is an acronym for *"centimetre, gramme, second"*. Refer to *"Tesla"*, where 1 Gauss = 10^{-4} Tesla.

GOES. Geostationary Operational Environmental Satellites. A fleet of satellites operated by the NOAA US National Weather Service monitoring both terrestrial and solar weather.

Hale's Law. Sunspot groups in a particular solar hemisphere will all be magnetically aligned latitudinally with the same polarity. The magnetic polarity in one hemisphere will be in opposition to the other. Proposed by US astronomer George Ellery Hale (1868 to 1938).

Helioseismology. A study of the Sun's internally generated acoustic pressure waves made by measuring Doppler shifts in the photospheric absorption spectra. The SOHO satellite Michelson Doppler Imager (MDI) measured shifts in the visible red around a Nickel emission line at 676.78 nanometres and the Global Oscillation and Low Frequency (GOLF) equipment monitored shifts in the sodium lines. By using very narrow band filters, shifts become apparent as changes in intensity, as the observed wavelength drifts off the filter peak and down its response curves in much the same manner as tuning in a radio station. The SDO mission has superseded the SOHO mission.

Hinode. A joint mission between the Japanese Aerospace and Exploration Agency, (JAXA), the United States and United Kingdom, to explore the Sun's magnetic fields. Launched in September 2006, the satellite employs optical, extreme ultraviolet and x-ray instrumentation.

Interplanetary Magnetic Field, IMF. The Sun's magnetic influence extends throughout the Solar System and beyond, carried by the solar wind. The field intensity changes as solar activity changes. Using magnetometers in ground stations, the influence that the IMF exerts onto our terrestrial magnetic field is measured by noting local changes, particularly in the direction of polarization.

Ion. An atom missing one or more electrons. It has a net positive charge equal to that carried by the missing electrons.

Isotope. An isotope is an atom of a chemical element containing more or less than, the regular number of neutrons.

Jeans' Criteria for the Collapse of a Nebula, Jeans' Instability Criteria. This criteria describes the conditions under which pockets of gas coalesce within a nebula and proceed to form stars and planets.

Joy's Law. The leading sunspot and its associated group will tend to incline towards the solar equator at an angle commensurate with the group's latitude. Proposed by US astronomer Arthur Harrison Joy (1882 to 1973).

Kelvin. A measurement of absolute temperature denoted by the letter "K", and equal to -273.16° Centigrade.

Magnetosphere. The region surrounding the Sun or a planet enclosing its own native magnetic field.

NASA. National Aeronautics and Space Administration. The US government's agency responsible for the US space programme, aeronautical and aerospace research.

Neutrino. A sub atomic elementary particle carrying no electrostatic charge, with small or no apparent rest mass.

Neutron. A sub atomic elementary particle, carrying no electrostatic charge, with a rest mass 1.674×10^{-27} kilogrammes.

Neutron Star.   A neutron star is the compressed stellar remnant left over from a supernova event, composed almost exclusively of neutrons. They are very hot, have a massive gravitational field and can spin at extreme rates.

NOAA. National Oceanic and Atmospheric Administration. A US government scientific agency within the Department of Commerce concerned with all aspects of solar and World weather.

Nuclear Fission. This is a process whereby the nucleus of a heavy element, such as uranium, is broken down into the nucleii of "less heavy" elements, such as barium, krypton and lead. The process is triggered by an input of energy in the form of an impacting neutron which breaks apart the atom's nucleus, resulting in the formation of lighter nucleii and the release of secondary neutrons. This is not to be confused with the similar process of supernova nucleosynthesis where temperatures and pressures of cataclysmic proportions prevail to create heavy elements like uranium in the first place.

Nuclear Fusion. This is a process whereby light atomic nucleii fuse with protons or other nucleii to form the nucleus of heavier elements whilst releasing energy in the form of gamma rays and sub atomic particles. This process creates elements up to iron, and possibly nickel, the two elements with the highest nucleon binding energy and mass deficits.

Perihelion. That point on Earth's elliptical orbit, at its nearest approach to the Sun.

Planetary K Index. Also called the "kp" index. It is a measure of the disturbance in the horizontal component of the Earth's geomagnetic field caused by the solar wind, used to predict the extent of auroral activity. Measurements are made using ground based magnetometers at a number of stations to provide a map showing the regional severity of the disturbance. Readings are published at 3 hour intervals. The level of disturbance in nanoteslas,"nT", (= 10^{-9} T), is tabulated on a scale of 0 to 9, where 0 represents a change between 0 to 5nT, and 9 is greater than 500nT. Typical daily fluctuations would be about 25nT. The ambient levels of the horizontal component of the Earth's geomagnetic field strength range from 30 microteslas, (= 30 x 10^{-6} T), around the equator to 60 microteslas at the poles.

Planetary Nebula. Neither a planet nor a nebula, but a small, white, hot star reaching the end of life. Nucleosynthesis has ceased, its outer atmosphere has blown away and its core collapsed under the force of its own gravity.

Plasma. The fourth state of matter. High energy gas composed of ions, protons and electrons, so hot that recombination cannot occur.

Positron. Antimatter counterpart of the electron, with the same mass of an electron, but with a positive charge of the same magnitude.

Proton. A positively charged sub atomic elementary particle, with a charge of 1.602 x 10^{-19} Coulombs, rest mass 1.672 x 10^{-27} kilogrammes.

Pyrheliometer. (Pyranometer). An instrument for measuring the solar constant. It is a type of thermopile sensor, a transducer which produces an output voltage proportional to incident radiation.

Red Giant, Red Supergiant. Classes of large, relatively cool stars, approaching the end of their life on the Main Sequence line in the Hertzprung-Russell Diagram.

RWC. The Belgian Regional Warning Centre at SIDC is one of thirteen worldwide, part of the International Space Environmental Service, ISES, operating through a NOAA hub in Boulder, Colorado, USA. They issue alerts when potentially damaging solar activity occurs.

Saha's Ionization Equation. Also called the Saha Langmuir Equation. Named after the Indian astrophysicist Meghmad Saha (1893 to 1956) and the US chemist and physicist Irving Langmuir (1881 to 1957). The equations describe how a plasma is formed from hot gases due to thermal collisions between the atoms knocking electrons out of orbit, creating ions, which then coexist with neutral atoms and free electrons. The equation describes the relationship between the gas temperature, density, and the degree of ionisation.

SDO. Solar Dynamics Observatory. A NASA solar observation satellite launched February 2011, and placed in geostationary orbit.

Sfu. Acronym for Solar Flux Unit. A measure of the field strength of solar radio emissions in the 10.7 centimetre waveband. Emissions on this wavelength correlate very well with sunspot counts and are used as a measure of solar activity. A solar flux unit equates both field strength and the emission frequency or wavelength in the following manner:---

$$Sfu = Watts/metre^2/nanometre,$$

where the 10.7 centimetre wavelength is 10.7×10^7 nanometres.

Expressed in terms of the frequency used, it is expressed as:---

$$Sfu = 10^{-22}Watts/metre^2/hertz$$

where the frequency (hertz) is that of the 10.7 centimetre waveband, 2.8 Gigahertz, = 2.8×10^9 hertz.

SIDC. An acronym for the Solar Influences Data Analysis Center. It is the solar physics research department of the Royal Observatory of Belgium, and the world data centre for the issue of the International Sunspot Number, ISN, and home to the Belgian Regional Warning Centre, RWC. It is the world centre for space weather forecasting and the issuance of solar storm warnings.

SOHO. An acronym for SOlar and Heliospheric Observatory. A joint ESA/NASA project satellite placed in geostationary orbit, to observe the Sun, launched December 1995, and planned to retire in December 2012.

Solar Constant. The solar constant is a measure of the flux density of all radiation, spanning the entire electromagnetic spectrum that reaches the Earth from the Sun.

Solar Cycle. This is an average 11 year cycle where the appearances of sunspots increase in number from a time of low solar activity to a period of maximum activity and back down again. The cycle ranges between 7 and 16 years. When a new cycle starts, the magnetic polarities of the solar hemispheres reverse, returning to their original state at the end of the following cycle. For this reason, some astronomers consider the average cycle length to be 22 years.

Solstice.   The summer solstice occurs in the Northern hemisphere around June 22, when the Sun ascends to its highest point in the sky and the Earth's pole of rotation inclines directly towards the Sun.   The winter solstice occurs around December 22 when the Sun attains its lowest altitude and the Earth's pole of rotation is inclined directly away from the Sun.   Dates vary due to the inclusion of Leap Years.   The opposite is the case in the Southern Hemisphere.

Spörer's Law.   Named after the German astronomer Gustav Spörer, (1822 to 1895).   It states that at the start of the solar cycle, sunspot groups tend to form at latitudes 30° to 40° either side of the equator.   As the cycle progresses, they form at lower latitudes migrating toward the solar equator.   Around the solar maximum they occur in the region of latitude 15° ending the solar cycle at latitudes 5° to 10°.

Stefan Boltzmann's Law.   The energy radiated from a black body surface over a period of time, called its radiant flux or irradiance, is proportional to its temperature.   The Sun is considered to act as a black body radiator with a temperature of 5776K.   It may be stated as the power emitted per square metre is proportional to the fourth power of its temperature, in Kelvin:

$$P = \sigma T^4, \quad \text{where } \sigma = 5.67 \times 10^{-8} \text{ Watts/metre}^2/K^4$$

Supernova. A supernova marks the death of a large star.   The star collapses in on itself under the force of its own gravity resulting in a very bright massive explosion, lasting for days or even weeks.   It is the only known phenomena in the Universe with sufficient energy to synthesise chemical elements beyond iron through to uranium.

SWPC. Space Weather Prediction Center. An organisation within the NOAA, part of the US National Weather Service. Its function is to provide solar weather reports and issue warnings of solar activity likely to have an effect on terrestrial operations.

Tesla. The SI unit of measure of magnetic flux density denoted by the letter "T". SI, Système International d'Unités, (International System of Units). Refer to *"Gauss"*, where 1 Gauss = 10^{-4} Tesla.

TRACE. An acronym for Transition Region And Coronal Explorer. Launched in April 1998 and retired in June 2010, a NASA space mission to explore the Sun's corona.

Van Allen Belts. There are two belts, or regions of radiation, stretching from pole to pole and encircling the Earth's equator, within the Earth's magnetosphere. They were named after the scientist James Van Allen whose equipment aboard the NASA Explorer 1 spacecraft discovered them in 1959. There are no definitive boundary limits to these regions and so data regarding their size varies from source to source. The inner belt is quoted as reaching from an altitude between 1,000 kilometres and 10,000 kilometres, centring on 6,400 kilometres, or about 4,000 miles above the equator, and contains high energy protons and alpha particles. The outer ring stretches between 13,000 kilometres to 60,000 kilometres, centred on 35,000 kilometres to 40,000 kilometres, around 25,000 miles, and is composed mainly of electrons with some protons.

White Dwarf. Class of relatively small, hot stars, in their declining years of life, having exhausted all their nuclear fuel with insufficient mass to become neutron stars.

Wien's Law.   Wien's Displacement Law states that the emission spectra of an object changes with temperature. Its temperature can be determined knowing the wavelength of the emission spectrum peak and applying a formula. Derived in 1896 by German physicist Wilhelm Wien (1864 to 1928).

Acknowledgements

The author wishes to thank the following people and organisations.

On-line encyclopaedia Wikipedia, for use of illustrations and images used in figures 1, 14, 29, 39, 44, 51, 52, 59, and 66.

The NASA SDO, STEREO, TRACE missions, the Goddard Space Flight Center and Marshal Space Flight Center, and joint missions ESA/NASA/SOHO and UK/US/JAXA Hinode for use of illustrations and images used in figures 8, 9, 10, 37, 38, 40 to 48, 50, 61, 62, 64, 65, 67 and 69.

The Institute for Solar Physics, the Royal Swedish Academy of Science, for use of the image used in figure 63.

The "solarstorms" website for information on the cost impact to United Airlines and of the damage to satellites due to solar storms.

A licence has been granted to use data from the article "The Composition of the Sun", published in the "Annual Review of Astronomy and Astrophysics", 2009.

---- to my wife Babs for her continued patience, understanding and help in producing this manuscript---

Useful Websites

http://sohowww.nascom.nasa.gov/data/realtime-images.html

Realtime images of sunspots and magnetograms from the NASA SDO mission, and archive images from the ESA/NASA SOHO mission.

www.atoptics.co.uk.

Les Cowley's free download TiltingSun program providing heliographic data.

www.petermeadows.com.

Peter Meadow's free download Helio program suite providing heliographic data.

www.ga.gov.au/geodesy/astro/sunrise.jsp

Australian government website providing sunrise and sunset, or moonrise and moonset data for any worldwide location.

www.solarstorms.org

A general interest program offering realtime and archive solar weather information.

www.spaceweather.com

A general interest program offering comprehensive realtime and archive astronomical and solar weather information.

<u>www.swpc.noaa.gov</u>

The NOAA Space Weather Prediction Center website, providing up to the minute solar weather information.

<u>www.swpc.noaa.gov/ftpmenu/forecasts/srs/html</u>

The NOAA Space Weather Prediction Center 24 hour solar region summary providing data on current sunspot groups with full classifications and heliographic coordinates.

Index

www.ingramcontent.com/pod-product-compliance
Lightning Source LLC
Chambersburg PA
CBHW031945170526
45157CB00002B/389